Electricity from Sunlight: The Future of Photovoltaics

Christopher Flavin

Worldwatch Paper 52
December 1982

Table of Contents

Introduction

olar photovoltaic cells have been called the ultimate energy technology. Environmentally benign and without moving parts, solar cells directly convert sunlight, the world's most abundant and widespread renewable energy source, into electricity, one of the most versatile and valuable forms of energy. Photovoltaic energy conversion is fundamentally different from all other forms of electricity generation. Without turbines, generators or other mechanical equipment, it more closely resembles photosynthesis, the biochemical process that forms the energy basis of life on earth.

Solar photovoltaics may become one of the most rapidly expanding energy sources—and one of the biggest growth industries—of the late twentieth century. Photovoltaics production has increased at a rate of more than 50 percent annually for the last five years, and a steady stream of companies is entering the solar electricity business. Many governments have dramatically boosted their support of photovoltaics and international competition is growing. Amid steep declines in the use of oil, recent abandonment of synthetic fuel projects and financial troubles in the nuclear power industry, photovoltaics is a striking exception: a healthy "sunrise" industry in a sea of economic and energy troubles.

A product of mid-twentieth century solid-state physics, photovoltaics does not fit the popular image of solar energy. It is a world of high technology laboratories, secret plans, patents and weekly rumors of the "latest breakthrough." Most of the pioneering research is done by specially-trained physicists and engineers who measure their progress in parts per billion, a dramatic contrast to the mega-engineering dominating the utility industry today.

Yet the promise of photovoltaics has attracted a wide range of supporters: back-to-nature homesteaders, environmentalists, politicians

I wish to thank Richard Curry, Robert Ferber, Calvin Fuller, Robert Johnson, Henry Kelly, Leonard Magid, Paul Maycock, Richard Munson, Alan Postlethwaite, Mortimer Prince, Jeffrey L. Smith, Barrett Stambler and Edward Sullivan for their reviews of the manuscript.

of all stripes and hardened businessmen. Increasingly, photovoltaics is a world of big business, with oil companies contributing a sizable share of the nearly half-billion dollars invested worldwide in 1982.

Although research and development still form much of the solar cell industry, practical use of the technology is growing rapidly. Many of us regularly rely on solar electricity when making long distance phone calls since all communications satellites are powered by photovoltaic cells. And in remote areas, telephones and other communications equipment often rely on photovoltaics. On a more limited basis, solar cells provide power for water pumping, desalination and lighting. And there are approximately 10,000 solar electric residences, mainly in rural areas without access to utility power lines. Ray Rudd, a retired welder who lives in the Beartooth Mountains in northern Montana with the help of solar electricity, notes that, "When you're used to no lights in the hills, it's pretty darn slick to have solar lights."[1]

These early uses for solar cells only hint at the ultimate worldwide potential of photovoltaics. Solar electricity systems can provide refrigeration or milled flour at remote Third World villages. When mounted on suburban rooftops, solar cells can supply reliable inflation-proof electricity, as well as independence from outside sources of energy. Photovoltaics can also be installed at large, centralized solar electricity stations operated by utilities, helping to diversify the energy sources feeding into power grids.

Photovoltaics is a proven technology with an impressive record of reliability. Yet the widespread use of solar cells hinges on one major challenge: cost reduction. At the current price of just under $10 per peak watt of generating capacity it would take a $50,000 photovoltaic system to meet the power needs of an energy-efficient modern home.[2] Prices have fallen by 50 percent in the last five years alone, however, and new technologies and manufacturing processes are at hand that virtually guarantee major cost reductions during the rest of the eighties. Since there are so many uses for photovoltaics, no single price makes it economically competitive. Already solar electricity is economical for communications systems and for many small village applications. Soon it will be used by remote industries and towns as well as by some utilities. Eventually photovoltaics will become a common rooftop technology throughout the world.

Approximately 60 companies are now manufacturing photovoltaics in 20 different countries. Already over half the solar cells produced each year are exported, and investment strategies and marketing skills are becoming as important to success in the industry as technological advances. Also crucial are the policies of governments, which have played a major role in the development of solar cells since the fifties. Severe reductions in the U.S. Government's photovoltaics program by the Reagan administration have been welcomed by European and Japanese competitors, whose governments are stepping up their support of photovoltaics. The United States is now in danger of losing its once overwhelming lead in the industry.

Developments planned or already under way make rapid progress in photovoltaics a near certainty in the next few years. Worldwide production of solar cells will expand at least fiftyfold and become a multi-billion dollar business by the early nineties. And sometime around the middle of the next century the world will likely get between 20 and 30 percent of its electricity from photovoltaics.[3]

Yet solar electricity will be even more important than these numbers indicate. Just as solar cells today serve an irreplaceable function aboard satellites, they will also meet many needs on earth that would otherwise go unfulfilled. For many among the more than one billion people still living without electricity, photovoltaics represents their best chance for a reliable power supply.

A Space Age Technology

In early 1954, a small team of scientists at Bell Laboratories in the United States was trying to find a practical way to generate electricity for telephone systems in rural areas not connected to a power grid. The Bell researchers were already familiar with the "photoelectric effect"—that light falling on certain materials could cause a spark of electricity—thanks to the work of French scientist Edmund Becquerel in 1839. Twentieth century physics, including the pioneering work of Albert Einstein, revealed that tiny "photons" or particles of light can strike the electrons surrounding atoms, causing a free stream of electrons—the basis of electricity. Using this knowledge, scientists developed primitive photovoltaic cells soon after the turn of the century. Made of selenium, these cells were extremely inefficient and exorbitantly expensive, little more than a scientific curiosity. Turning this idea into a practical power source was a daunting challenge indeed.[4]

As often happens in such endeavors, the initially unsuccessful efforts of the Bell researchers were transformed by a bit of serendipity. While the scientists were trying to improve the efficiency of selenium cells, a separate Bell research team was developing an electronic device made of crystalline silicon. This team accidentally discovered that the silicon device produced electricity when exposed to sunlight. The scientists immediately realized that the two research efforts should be merged, and within days they achieved a remarkable breakthrough. Whereas the selenium cells converted only 1 percent of the sunlight they absorbed into useful electricity, the silicon solar cells converted eight percent, a dramatic increase in efficiency. The scientists had another reason for excitement. The material they were using—silicon—is the world's second most abundant element, composing 28 percent of the earth's crust.

These research achievements were greeted with much fanfare amid the general technological euphoria of the fifties. Solar cells seemed to promise an unlimited supply of electricity, and the news media jumped briefly on the photovoltaics bandwagon. In 1957 *Business Week*, celebrating advances in photovoltaics research, envisioned an automatically controlled solar car in which "all the riders could sit comfortably in the back seat and perhaps watch solar-powered TV."[5]

It was an inauspicious time, however, to develop a new energy technology. Oil cost less than $2 a barrel and large fossil fuel power plants were being built at a record pace. The price of electricity was falling. Moreover, 1954 was the same year that construction began on the

"It was the space program—the very
symbol of modern technology—that
rescued photovoltaics from the
technological scrap heap."

world's first commercial nuclear reactor. Nuclear power was going to
be the source of electricity "too cheap to meter." Massive government
programs had been launched to harness the atom. Compared with
these efforts the idea of sunlight as an energy source seemed primitive.[6]

Photovoltaics researchers also faced an unsettling economic reality.
Silicon cells developed in the fifties were prohibitively expensive,
with costs as high as $600 per watt or $600,000 per kilowatt, several
hundred times as high as conventional electricity sources.[7] After
building a few prototype photovoltaic panels, Bell Laboratories shel-
ved its development efforts. Although cost reductions were clearly
possible, photovoltaics as a competitive, practical electricity source
appeared to be well over the horizon. The major research effort
needed to lower costs could not be funded in an era of declining
electricity prices and booming nuclear power plant construction.

It was the space program—the very symbol of modern technology—
that rescued photovoltaics from the technological scrap heap. American scientists in the late fifties were searching for a lightweight and
long-lasting power source for satellites. Photovoltaic cells were the
answer. They have the additional advantage of being more efficient in
the 24-hour intense sunlight available in space. In 1958, just four
years after the Bell Laboratories breakthrough, silicon solar cells were
boosted into orbit aboard Vanguard I, the second U.S. satellite.[8]

With the help of large contracts from the National Aeronautics and
Space Administration (NASA), four U.S. companies entered the
photovoltaics business and by the late sixties were producing hundreds of thousands of solar cells a year (enough to generate 80 kilowatts of power). Amid the heady competition of the post-Sputnik
space race, the Soviet Union was soon equipping its satellites with
photovoltaics as well. Today solar cells power virtually all satellites,
including those for defense as well as scientific research. Solar electricity is particularly important to the growing world information
economy since solar cells are used on satellites that relay long distance telephone calls, computer hookups and television trans-
missions.

9

The achievements in solar cell research during the peak years of the space program included a major increase in efficiency and a modest reduction in cost to between $100 and $200 per watt by 1970.[9] Yet photovoltaics was still far from practical for terrestrial use. Much of the space program's research efforts improved the efficiency and durability of solar cells exposed to the rigors of conditions beyond the atmosphere. In the early seventies solar cells were still at least 50 times as expensive as conventional technologies for electrical generation. But cost mattered little since relatively few cells were required and the NASA budget was astronomical anyway.

Rising fossil fuel prices in the seventies spurred researchers to develop solar cells for use on earth. Between 1972 and 1975 new research programs began in Europe, Japan, the Soviet Union and the United States. Perhaps a dozen private companies entered the solar cell research or production business. Although substantial technological challenges remained and significant profits in photovoltaics were at least several years away, the potential rewards seemed well worth the risks to many scientists and entrepreneurs.

Major advances have been made in solar cell technology in the last decade, boosting the efficiency of commercial photovoltaic modules to 10 percent and higher while improving reliability. More importantly, solar cell costs have fallen to a tenth their level in the early seventies. As a result, a steadily growing market for solar cells has been established and new companies have continued to enter the business. Worldwide production of photovoltaics is measured by the kilowatts of peak capacity that the solar cells can generate. The annual manufacture of photovoltaics has grown from a few hundred kilowatts of peak power capacity in the mid-seventies to about 8,000 kilowatts (8 megawatts) in 1982—100 times the highest space program level. (See Figure 1.) Altogether the solar cells manufactured in 1982 can provide as much power as is used by approximately 1,800 energy-efficient modern houses.[10]

Single-crystal silicon cells similar to those produced by Bell Laboratories continue to dominate the industry, although many othr types of solar cells are being developed. Silicon is one of a family of ele-

**Megawatts
of Capacity**

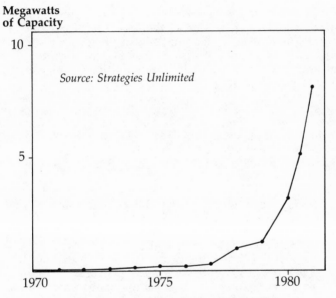

Source: Strategies Unlimited

**Figure 1: Annual World Production of
Photovoltaic Cells, 1970-1982**

ments known as semiconductors that have properties in between
those of a metal and nonmetal and so conduct electricity only slightly.
This is the basis of silicon's photovoltaic properties, but several steps
are required to turn ordinary silicon—the main component of sand—
into a solar cell. The silicon must first be separated from the oxygen
with which it is normally combined and then purified to the point
where there is less than one nonsilicon atom per billion. This semiconductor grade silicon is one of the purest commercial materials in
the world, with a price tag between $60 and $100 per kilogram.[11]

The purified silicon is melted at a very high temperature and then
slowly drawn from a vat to form a large crystal about 15 centimeters

in diameter and as much as a meter long, a procedure known as the Czochralski process. The crystal is then laid on its side and cut into thin round wafers using a diamond saw, a procedure resembling an extremely difficult version of slicing bologna. Unlike bologna, half of the silicon crystal, now worth hundreds of dollars per kilogram, is turned into useless dust by the slicing process.

The thin slices of silicon crystal that remain must contain traces of two materials in order to generate electricity. One, usually boron, is added to the molten silicon before it is crystallized. Phosphorous is later added to one side of each wafer, forming a barrier of electric charge between the two parts of the solar cell that directs the flow of electrons.

Metal contacts placed on the front and back of the solar cell form an electric circuit, and the cell is sealed with soft plastic and glass for protection. Groups of cells are mounted on a rigid plate and wired together to form a photovoltaic panel or module, typically a square meter in size, with a generating capacity of approximately 100 watts. The whole device resembles a water heating solar collector except for its distinctive dark round cells. When sunlight hits a solar module the silicon electrons leave their atomic "shells" and, because of the electronic patterns of the photovoltaic device, electrons flow through wires that join the sides of the cells. Today commercial solar modules typically turn 10 to 12 percent of the sunlight that strikes them into electricity, and laboratory efficiencies of 15 percent and higher are common.

Several hundred companies are now involved in the photovoltaics industry worldwide. Of these about 60 manufacture and sell photovoltaic modules and another hundred produce raw materials, components or solar electric products. A much larger group of companies, numbering well over a hundred, and including some major corporations, conducts research and development on new types of solar cells. Some of these companies may soon enter the commercial market with new photovoltaic technologies. Total sales of solar electric systems climbed to approximately $150 million in 1982, and private investment has reached close to $300 million annually, another indication that the industry is moving from infancy to adolescence. France, Italy, Japan, the United States and West Germany are the leading countries in photovoltaics today.[12].

13

Current uses for solar cells are quite diverse, though they have proved most popular in applications for which conventional energy sources are impractical or simply not available. About a third of today's solar electric systems are used at remote communications installations. Approximately 20 percent provide power for water pumping, and another 15 percent are used at isolated houses. The remaining solar cells meet needs ranging from electrical rust protection of bridges to the lighting of fire lookouts and aviation beacons. Most of these systems are still expensive, particularly since they usually require batteries to store electricity for times when the sun is not shining. But they are usually less expensive than any other technologies that can be used for such small-scale remote applications.[13]

This remote-applications market for solar electric systems continues to grow rapidly. Today oil companies use photovoltaic power to help monitor the operation of their remote pipelines. Papua New Guinea, one of the least developed parts of the world, has a sophisticated telecommunications network run by solar electricity. Photovoltaics is also a main power source for pumping water, refrigeration and lighting in Papua New Guinea. And photovoltaic systems are literally saving lives in the Canadian Rockies and Swiss Alps, where solar electricity is used at remote rescue stations for stranded mountain climbers and cross-country skiers.[14]

Photovoltaics has already established an impressive record of reliability in extraordinarily harsh conditions. In space solar cells work in a vacuum at extremely high and low temperatures while constantly exposed to intense radiation. Back on earth photovoltaic systems have been installed in communications systems on the frozen tundra, in fire lookouts atop high peaks and with water pumps in parched deserts. Tougher tests for a new technology would be hard to design, but there have been remarkably few solar cell failures. In fact, the first photovoltaic system sent into space operated without interruption for eight years—needlessly occupying a radio band since its transmitter had no cutoff device.

14 Solar cells are reliable because they are solid-state devices with no moving parts and relatively little to go wrong. Sunlight interacts only with the electrons in the cell and does not alter the material itself, so solar cells can last indefinitely. This is not true of any technology that has mechanical parts—which must eventually wear out—such as a wind turbine, diesel engine or nuclear power plant. The lifetime of photovoltaic systems is limited only by supporting and protective equipment that will likely last from 20 to 30 years. Reliable systems, however, must be securely enclosed in plastic and glass since water and other impurities can cause corrosion. Removing accumulated dust or occasionally replacing storage batteries that may be included in the system is the only maintenance required.[15]

The Cost Reduction Game

Cost reduction is the key to solar electricity's current role and future. Photovoltaics has so far followed a remarkable cost curve, with prices falling by an average of more than 50 percent every five years. Recently, average solar module prices have plummeted from $20 per peak watt in 1977 to less than $10 in 1982. (See Figure 2.) In few other industries have costs fallen so rapidly. Yet even at today's prices, a photovoltaic system just large enough to power a 100-watt light bulb would cost at least $1,500, so further cost reduction is obviously essential.[16]

R&D programs to reduce costs continue to take precedence over marketing and sales in the photovoltaics industry today. Some industry leaders believe it is too soon to market photovoltaics and that commercialization could distract from efforts to improve technologies for the much larger markets of tomorrow. Even companies already selling millions of dollars worth of solar electric systems each year are plowing revenues back into research and development in order to keep pace as the industry evolves. All realize that photovoltaics is a dynamic field in which standing still could mean falling behind.

"Solar cells are reliable because they are solid-state devices with no moving parts and little to go wrong."

U.S. Dollars per Peak Watt

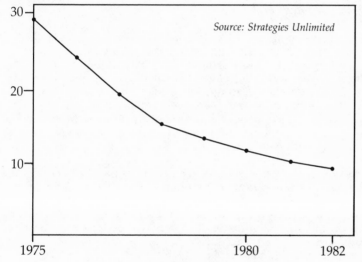

Source: Strategies Unlimited

Figure 2: Average Market Prices of Photovoltaic Modules, 1975-1982

The largest industrial R&D programs in photovoltaics are in those countries that entered the field first: France, Italy, Japan, West Germany and the United States. Other countries with smaller programs include Australia, Belgium, Brazil, Canada, China, England, India, Mexico, the Netherlands, the Soviet Union, Spain and Sweden. No other country can yet match the size and diversity of U.S. R&D efforts, but the Europeans and Japanese are challenging for the lead in some areas of research.[17]

In all these countries government plays a major role in developing photovoltaics, often channeling R&D funds to universities as well as private companies. And in most nations, government assistance for photovoltaics has grown. The $150 million spent on solar cell technology by American, European and Japanese governments in 1982 made it the second most funded renewable energy technology, just behind solar heating and cooling.[18]

Private investment in solar cell technology has risen even faster. Whereas in the mid-seventies government R&D dwarfed private programs, approximately two thirds of the $500 million spent in 1982 came from private industry. Governments direct much of their support to basic research and long-term goals, while private firms emphasize short-term aims and the development of commercial technologies. Increasingly, private investment in photovoltaics is being used to develop assembly line manufacturing processes that can greatly lower production costs.

Both government and private research programs have set specific cost goals for each of the various components of a system, from silicon refining to final installation. These programs regularly monitor progress in the various areas and occasionally redirect research efforts to more promising materials or designs. The goals are undeniably ambitious, aiming to cut prices five- to tenfold over the next decade. These targets would have seemed fanciful in an era when technology evolved at a more leisurely pace, but this is the age of synthetic materials, microelectronics and sudden breakthroughs. Photovoltaics and microelectronics, in fact, are close cousins since they both rely on semiconductors—one reason that researchers are optimistic that cost goals can be met.[19]

The current cost of photovoltaics is high for a number of reasons: expensive materials, energy-intensive processing, large labor requirements (including the meticulous hand-assembly of cells), and extensive supporting equipment such as the array structure, electrical wiring and battery storage. Three broad avenues to reducing costs are being followed. One is the development of inexpensive automated methods of manufacturing single-crystal silicon cells—the dominant technology since the mid-fifties. The second is to develop any of several alternative solar cells that are likely to be much cheaper to produce. The main challenge here is to obtain acceptable efficiencies of at least 8 to 10 percent using solar cell materials without the in-

"The goals are undeniably ambitious,
aiming to cut prices five- to tenfold over
the next decade."

herently good photovoltaic properties of crystalline silicon. In a third cost reduction strategy, researchers are designing systems that greatly concentrate sunlight and so lower the cost of photovoltaic electricity, even with today's expensive solar cells.

17

The largest and most heavily funded R&D effort is to upgrade and automate the manufacture of single-crystal silicon cells. By one count, over a hundred industry and university research teams are engaged in various aspects of this effort. Researchers aim to cut the production costs of purified silicon by at least two-thirds, to develop high-speed crystal pullers and high-speed, multiple-blade slicing for the manufacture of the wafers, and to automate the cell and array assembly processes. Some of these improvements have already been introduced in commercial plants and others will be in the next few years. Scientists believe that simply automating the manufacture of single-crystal silicon cells using recently developed technologies could cut the cost of photovoltaics by more than half within the next five years, while raising efficiency to at least 15 percent. [20]

Meanwhile, more innovative approaches to making low-cost photovoltaics are also under way. Several companies in the United States are directly producing large sheets of nearly single-crystal silicon using continuous ribbon growth techniques. These bypass the expensive and wasteful crystal growing and slicing stages and are a major step toward automation. So far ribbon growth processes have succeeded in the laboratory, yielding high-quality efficient cells, that compare favorably with the performance of conventional single-crystal cells, and small-scale commercial production is under way. The next step is large-scale commercial production, which may happen by the mid-eighties. This would be a significant advance for the industry and probably provide competition for conventional silicon cells. [21]

Another solar cell technology with considerable potential is the polycrystalline silicon cell (composed of many crystals). Sliced from a large silicon ingot that is produced through an inexpensive casting process, these cells can be made from a less pure form of silicon that costs only a fifth as much as semiconductor-grade silicon does today.

The main challenge is in attaining a sufficiently high efficiency since the "grain boundaries" between the crystals naturally reduce the effectiveness of a photovoltaic cell. Industry leaders are confident, however, that polycrystalline solar modules can achieve 8 to 10 percent efficiencies at a competitive price. Two companies in the United States and one in West Germany are independently developing polycrystalline photovoltaics. One U.S. company began manufacturing the devices commercially in 1982 and soon will produce its own "solar grade" silicon for these cells.[22]

Receiving even more research attention are the so-called "thin-film" solar cells made from a variety of inexpensive materials, including amorphous silicon and cadmium sulfide. Researchers hope that these materials can one day simply be spray-coated on glass or rolled out in a continuous process much the same way photographic film is today. Because such materials can be a few micrometers or less in thickness (less than a tenth as thick as conventional photovoltaics), materials costs can be reduced greatly. Photovoltaics specialist J. Richard Burke believes that "the low-cost pot at the end of the rainbow lies in the use of truly thin-film photovoltaic cells."[23] Indeed, if the currently modest efficiencies of such cells can be raised by 30 percent or so and production costs drop as low as expected, thin-film technologies could swiftly transform the photovoltaics market. Yet, thin-film solar cells have skeptics as well. Just how they work is still not completely understood, and some thin-film materials may degrade over time. Many researchers also question whether manufacturing processes for efficient thin-film photovoltaics are likely to be commercialized in the near future.

Solar cells made of amorphous silicon are being aggressively developed by industry and government in both the United States and Japan. Composed of randomly arranged or noncrystallized atoms, amorphous silicon is a complex material resembling ordinary glass, but with an unusual characteristic—it exhibits good photovoltaic properties with the addition of small amounts of hydrogen. Since today's amorphous silicon photovoltaic modules are only 3 to 6 percent efficient, improved efficiency has so far been the main target of research.

The U.S. industry has developed the most advanced amorphous silicon cells, but has chosen to concentrate on R&D. Meanwhile, beginning in 1980 Japanese companies blazed the way to a commercial market by installing the cheap, low-efficiency solar cells in pocket calculators where power needs are low and efficiency is relatively unimportant. This ingenious idea has allowed Japanese firms to start building large plants and so gain experience in perfecting the manufacturing processes for amorphous silicon. One result is that several U.S. firms that had been biding their time may soon begin selling amorphous silicon cells commercially.[24]

It probably will be close to a decade before amorphous silicon becomes a significant competitor in most markets. But its long-run potential to revolutionize the industry is enormous, mainly because amorphous silicon is ideal for high-speed, low-cost production. Many photovoltaics researchers think that it is the dream technology of the nineties.

Other materials that can be made into effective thin-film solar cells include cadmium sulfide, copper sulfide and cadmium telluride. Photovoltaics using these materials have not yet been marketed on a significant scale, but a great deal of research is under way and at least one U.S. company is designing a large cadmium sulfide production plant that it believes could yield competitive solar cells. Other researchers, however, think that cadmium sulfide is too unstable to make effective photovoltaics anytime soon. Additional thin-film materials are being investigated intensively in industry and university laboratories. Although most probably will never be widely used, surprising advances continue to be made regularly and the sheer breadth of the research efforts provides ample reason for optimism. One limitation on some of these solar cell materials, however, is that they are quite rare or poisonous—as is the case with indium or arsenic.[25]

Scientists are also at work on a number of innovative solar cell designs. Because particular photovoltaic materials vary in their sensitivity to different wavelengths of light, researchers are attempting to increase efficiency by stacking several different cell types and so convert a broader spectrum of the sun's radiation into electricity. This design, known as the "tandem cell" or "multi-junction cell," has already produced efficiencies as high as 25 percent in the laboratory, but its practical market potential is still uncertain. One intriguing idea is that thin-film and tandem cell designs might one day be combined, yielding a low-cost *and* high efficiency solar cell.[26]

Another exciting innovation is the electrochemical solar cell that uses sunlight to produce hydrogen or some other useful fuel rather than electricity. By mounting silicon or another semiconductor next to a special liquid electrolyte, the sun's energy can be used to separate hydrogen atoms from water molecules. Efficiencies as high as 13 percent have been achieved in the laboratory, and researchers are now developing a production process as well as seeking cheaper materials for these cells. Since hydrogen has considerable potential as a transportation fuel that can be substituted for gasoline, producing hydrogen inexpensively using the sun's energy would be a major step toward resolving the world's oil problems.[27]

Another approach to harnessing inexpensive solar electricity is an optical concentrator system that acts like a magnifying glass, increasing the amount of sunlight striking a solar panel 10 to a thousand times. The efficiency of most solar cells actually improves in more intense sunlight as long as they are kept cool. Therefore, inexpensive concentrators can help generate electricity economically even if the cells themselves are fairly expensive.

Parabolic troughs and Fresnel lens concentrators have received the most research attention in the last few years, and each design has backers who feel it can be used to cover wide areas at low cost—allowing farms, industries or utilities to generate power. Usually these systems include mechanical "trackers" that maintain an optimal angle to the sun throughout the day. The other key to a successful concentrator system is high efficiency cells—usually above 15 percent. Most concentrator systems employ either single-crystal silicon or gallium arsenide. The latter material has been used to produce solar cells with efficiencies of over 20 percent in laboratories in the

> "Researchers well-versed in the field say
> that anyone claiming to know what will
> be tomorrow's cheapest solar cell
> technology is a charlatan."

United States and Japan. So far the U.S. is clearly the leader in developing concentrator systems for use with photovoltaics.[28]

Photovoltaic concentrator systems face a number of problems that could limit their usefulness compared to simple "flat plate" solar systems. First, concentrators must include cooling systems to radiate away the heat that naturally builds up and reduces efficiency. One innovative solution turns this problem into an asset, using the excess heat to warm air or water. These experimental thermal-photovoltaic systems are in effect photovoltaic cogenerators. They have an overall conversion efficiency as high as 30 to 35 percent, but their practicality and cost-effectiveness have yet to be demonstrated. A second problem is that most concentrator systems include mechanical devices. They tend to break down more than nontracking systems and require more maintenance. A broader limitation on all concentrator systems is that they perform well only in direct sunlight, whereas flat plate photovoltaics work adequately under light clouds or haze. For this reason concentrator systems may prove popular only in sunny regions such as the Middle East, southern Europe, sub-Saharan Africa or the southwestern United States.[29]

Photovoltaic technology is becoming ever more complicated, offering a bewildering array of challenges and potentials. Researchers well-versed in the field say that anyone claiming to know what will be tomorrow's cheapest solar cell technology is a charlatan. Yet the diversity is what makes photovoltaics so promising. Primary materials, manufacturing techniques, array designs and systems technologies can be combined and recombined in dozens of ways. Major cost reductions during this decade are virtually guaranteed and do not require the spectacular breakthroughs achieved in the past.

Both government and private research programs have made major efforts to predict future costs and establish realistic goals, since cost is the benchmark against which this technology is judged. The most commonly cited goals are those established by the U.S. Department of Energy (DOE) in the late seventies. DOE aimed to develop the technology needed to bring market prices for photovoltaic modules down to $2.80 per peak watt (in 1980 dollars) by 1983, to 70¢ per watt by 1987 and 40¢ per watt by 1990. (See Figure 3.) The idea was to make photovoltaics fully competitive with conventional sources of electricity by the nineties.[30]

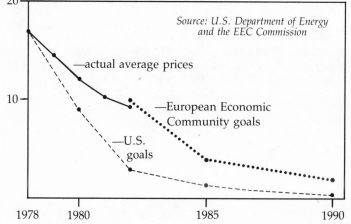

Figure 3: **Market Prices and Price Goals for Photovoltaic Modules, 1978-1990**

In retrospect, these goals were ambitious considering the modest R&D programs established to achieve them. Although technologies have been developed at about the pace expected, there are many steps before a laboratory technology becomes an assembly line product selling for just a few dollars per square foot. Laboratory efficiencies of 15 to 20 percent are no longer rare. But efficiencies of 10 or 12 percent in mass-produced inexpensive modules are proving much more difficult. Progress has been gradual rather than sudden, mainly because industry lacks the capital and large markets needed to support investment in large-scale manufacturing plants. More modest but still optimistic goals have been established recently by the

European Economic Commission and the Japanese government. Each calls for a 1985 figure of about $5 per peak watt (1980 dollars) and $2 per watt by 1990. Independent analysts concur with these projections and most see rapid market growth as a result of declining costs.[31]

In addition to lowered solar cell costs, the cost of several other components of a photovoltaic system must also be reduced. These components typically include electrical wiring, power conditioning, control devices, battery storage, support equipment for the solar modules, site preparation and installation. Collectively these are known as the "balance-of-system" costs and they typically run between $6 and $12 per peak watt, accounting for as much as 50 percent of total cost. As a result, a full solar electric system today will usually cost between $15 and $25 per watt and sometimes more, depending on the size of the system, whether batteries are required, and other variables. Balance-of-system costs have been falling in recent years but not nearly as rapidly as solar cell costs. This trend is expected to continue, and the balance-of-system could account for as much as 80 percent of the total cost by the early nineties.[32]

Lowering balance-of-system costs is quite different from lowering solar cell costs. Most of the technologies and procedures are already used extensively by other industries and are not likely to undergo the vast technical improvements that have characterized photovoltaics. Lead acid batteries that have been used in automobiles for over a half century, for example, are still the least expensive way of storing solar power and often account for well over half of the balance-of-system costs. Research and development on improved batteries is continuing, but improvements are marginal and commercialization of a new battery technology appropriate for photovoltaic systems is unlikely for at least ten years. Meanwhile, the price of batteries is actually rising because of the high costs of labor and lead. The costs of major materials used in solar electric installations—such as structural steel, concrete and copper wire—are even less likely to fall dramatically.[33]

Efforts are ongoing to improve power conditioners, cut support structure costs, and simplify wiring and installation procedures. Standardized photovoltaic systems and factory assembly are the most promising ways to lower these costs, but each requires a major investment. So far, the balance-of-system has been the black sheep of the photovoltaics family, and neither government nor industry is making a large effort to reduce these costs. It is likely, however, that as the market for photovoltaics grows industry will begin developing cheaper inverters, simpler array designs and the like. Meanwhile, governments should keep track of balance-of-system costs and provide targeted support for the industry in this area.

Although uncertainties remain, clearer price forecasts are possible than those a few years ago. Larger manufacturing plants employing more advanced and less expensive processes are scheduled to come on line in the next few years. And intense competition for market shares will tend to push prices down. Conventional crystalline silicon cells, together with ribbon growth and polycrystalline silicon cells, will likely dominate the market for the rest of this decade, though analysts differ as to which of these will be the most successful. Concentrators will probably be widely used in many applications, particularly utility plants. Beyond 1990 amorphous silicon and other thin-film technologies likely will capture the largest share of the market, pushing prices to new lows.

The photovoltaics market will evolve gradually rather than in discrete stages and at each point there should be a range of technologies to choose from—each with its specialized applications. Module prices will probably fall to approximately $3 per watt (1980 dollars) by 1987 and to about $2 per watt by 1990. At that price a total solar electric system will cost between $4 and $8 per watt and generate electricity at a cost of 15¢ to 30¢ per kilowatt-hour (as opposed to over $1 per kilowatt-hour today). This is getting close to standard electricity prices in many parts of the world, including Europe and Japan. Predictions beyond the early nineties are difficult to make since they are dependent on technologies barely beyond the laboratory stage. But further substantial cost reductions are likely since the photovoltaics market will be large enough to attract big investments. Given the rising costs of most sources of electricity, including coal and nuclear power, photovoltaics is likely to be a competitive electricity source in all but a few areas of the world by the mid-nineties.[34]

The Birth of an Industry

The photovoltaics industry has always stood apart from other renewable energy industries. Whereas wind power and solar water heaters first found support among environmentalists and do-it-yourselfers, photovoltaics has been a realm of three-piece suits and carefully crafted investment plans. Today the solar cell industry attracts over $300 million of private investment each year, perhaps two-thirds of it in the U.S. Sales have grown steadily to approximately $150 million in 1982.[35] The industry is diverse, with entrepreneurial firms of a dozen employees competing alongside electronics conglomerates and oil companies that have gained their position by gobbling up small firms "Pac Man" style. Which photovoltaics companies will succeed is uncertain, but clearly the strength of the industry will help determine how soon solar electricity is widely used.

Most of the early investment in photovoltaics in the seventies came from private risk capital and government R&D programs. As often happens in new industries with low entry costs and promising futures, small companies abounded, each pursuing a particular technology. In the United States the solar cell industry closely followed the growth pattern of the semiconductor industry a decade earlier: young scientists employed by large companies recognized the potential of photovoltaics and broke away to start their own firms. R&D funds came mostly from venture capital supplied by wealthy individuals who used the investments as tax writeoffs.[36]

Large electronics firms and other diversified corporations have also had a prominent role in photovoltaics. These companies used their expertise in related fields to build research teams and develop improved photovoltaic devices, particularly in Europe and Japan, and to a lesser extent in the United States. Only a few such companies have begun marketing significant numbers of solar cells, but many have impressive R&D programs and may be ready to commercialize advanced technologies in the near future.

The solar cell industry has expanded rapidly in recent years, and important qualitative changes are inevitable with this growth. Because of the depth and diversity of the technology, no entrepreneur can be assured that any given approach will succeed. An expensive and technologically successful R&D program could be wiped out overnight if a competitor achieves a dramatic breakthrough. The level of risk is rising because cost reductions increasingly depend on large-scale manufacturing processes that require sizable initial investments. Each year the amount of investment needed to attain a competitive position in the industry grows. Companies must somehow bridge the financial chasm that separates today's modest sales and R&D programs from the huge, virtually unlimited markets of tomorrow.

The largest source of new capital in photovoltaics in recent years has been the oil industry. By 1981, eight oil companies in the United States and two in France had invested in photovoltaics, together bringing over $100 million to the industry each year. Petroleum firms have purchased controlling interest in some solar cell companies, and they own a minority share in others. Three solar cell firms that are wholly or partly owned by oil companies—all from the United States—had an estimated 80 percent of the worldwide market for photovoltaics in 1981 according to the Center for Renewable Resources, a U.S. public interest group. Others hold a dominant position in developing new technologies not yet commercialized.[37]

The rush to invest in photovoltaics occurred simultaneously with the withdrawal by some oil companies from the solar water heater market. One explanation is that the solar collector industry depends largely on plumbing skills and the success of small retail outlets, areas where the oil industry has little experience. Although they also lack expertise in semiconductor devices, oil firms do have experience with long-term, high-payoff R&D programs. Because of generally tight capital markets and high interest rates, few other companies can invest tens of millions of dollars in a speculative industry where large payoffs are at least five or ten years away. Oil company executives see

"Oil company executives see photovoltaics as an emerging industry in
which they can use their financial
muscle productively."

photovoltaics as an emerging industry in which they can use their financial muscle productively. Oil industry strategists also see investment in photovoltaics as an important step in the creation of diversified energy companies that will outlast the petroleum era.

27

These developments have raised a red flag for many solar energy advocates. They fear that oil companies, with large holdings in natural gas, coal and uranium, are now getting a lock on sunshine as well. A 1980 book by Ray Reece entitled *The Sun Betrayed* alleges that oil firms intend to slow down the development of photovoltaics until oil wells run dry, when they can then monopolize the solar cell industry. Karl Frieden of the National Center for Economic Alternatives echoes these fears, asserting that "the oil companies will hold solar power hostage, while maximizing their profits from more limited conventional fuel supplies." A study of the photovoltaics industry by Barrett Stambler of the Center for Renewable Resources, though more restrained, concludes that the solar cell industry already lacks competitiveness and suggests that at least one company practices predatory pricing—selling photovoltaics below cost in order to drive other firms out of business.[38]

Most solar industry executives and many independent analysts reject these charges. George Tenet of the Solar Energy Industries Association, which has several oil company members, claims that, "The number of competitors has consistently increased and the selling price has consistently decreased . . . There's no attempt to stifle competition or suppress new technology." Tenet and others point out that oil and photovoltaics compete in separate markets and that too many companies are currently involved for any one firm to contain photovoltaics until the world runs out of oil. Industry analysts note that patent protection in photovoltaic technology is inherently weak and not likely to block the entry of new firms. And Morris Adelman, an economist at the Massachusetts Institute of Technology, concludes that "The notion that the energy giants, controlling the biggest part of the manufacturing capacity in photovoltaics, could set the price artificially high to protect their other investments is unrealistic."[39]

The truth in this controversy probably lies somewhere between the two extremes. Oil company investments have, in fact, spurred the photovoltaics industry, and so far the companies with oil industry support appear to have innovative, aggressive development programs. There is nonetheless cause for concern. Large corporations tend to be relatively hidebound and unimaginative, whereas smaller companies have made a disproportionate share of the world's industrial breakthroughs. In photovoltaics, the pioneering small companies have been bought out at an earlier stage in their development than was the case in semiconductors and most other technologies. Oil companies bring abundant capital to the solar cell industry, but they have no experience in this field and their record of innovation may not match that of smaller companies.[40]

The trend toward control of most energy sources by a handful of multinational corporations is itself cause for concern. Although the photovoltaics industry remains competitive today, large corporate control could damage its dynamism at some point in the future. Government programs deserve part of the blame since they have channeled most of their funds to larger, more established firms. Governments could foster more competition in the solar cell industry by making special loans and R&D grants to small, innovative firms most in need of capital.[41] Investment and innovation must go together if solar electricity is to be widely used in the near future.

In international markets competition in photovoltaics is virtually assured. Of the approximately 60 established solar module manufacturers, Europe has 12, the United States has 25 and Japan has 15. Another 8 to 12 companies are producing photovoltaics in developing countries, all on a small scale. Although U.S. firms still have the most advanced solar cell technology, their share of the market fell from 80 percent in 1980 to 55 percent in 1982. Competition is intensifying, and the number of companies continues to grow rapidly, particularly in the Third World via joint ventures with European and American companies. Obviously, early leadership will not necessarily translate into lasting strength in this rapidly evolving industry.[42]

The Europeans and Japanese are moving rapidly to shed their under-dog image. Firms in these countries have made significant progress in recent years, swiftly closing the gap in some fields. French photo-voltaics expertise has been consolidated under two major companies—Photowatt and France-Photon. The first has oil company and government funding and the second is a subsidiary of a U.S. firm. Similarly, West Germany has two large electronics companies dominating the solar cell business. Italy too has made a major com-mitment, with its government directly owning most of the country's growing photovoltaics industry. In Japan, large integrated electronics firms such as Fuji, Sanyo and Sharp have staked out a strong posi-tion. By studying American technology and carefully plotting their own course, Japanese companies have leapfrogged close to the "cut-ting edge" in amorphous silicon production and other areas, creating a stir in the photovoltaics industry around the world.[43]

Photovoltaics is developing in a climate of global information ex-change and international business relationships that would have been inconceivable a decade or two ago. Extensive travel and com-munication among scientists rapidly spread new technologies around the world, and production processes are flexible enough that patents cannot easily protect particular technologies. In the last few years, international joint ventures and licensing arrangements have become popular in photovoltaics, allowing companies to enter foreign mar-kets and boost sales. Solarex, an American firm that is one of the two largest in the world, has already established nine joint ventures, most of them in Europe. Japanese companies have signed licensing agree-ments with U.S. firms that have helped move Japanese researchers ahead in some technologies.[44]

Despite increased cooperation between some companies, inter-national competition in photovoltaics is intensifying. Solar cell pro-duction and the share of solar cell exports have both increased. Com-petition in the Third World—particularly the rapidly industrializing "middle-income" countries such as Brazil, Mexico and the Philippines—is already intense and may help shape the immediate future of the solar cell industry. International marketing skills will soon be as important as technical expertise in determining market share.

Experience indicates that European and Japanese industries will have some advantages in the market now evolving in photovoltaics. Firms in these nations have traditionally made over half of their sales abroad, and their executives have long courted Third World buyers. Indeed, some European and Japanese firms already market in Third World countries many of the technologies compatible with photovoltaics, such as pumps, generators and batteries. Establishing these channels takes years and is almost as difficult as developing a new technology.

Integrating solar electric systems into such existing markets already has given French firms an advantage in West Africa, West German firms an edge in Latin America, and Japanese firms an advantage in Southeast Asia. Some American companies have begun selling their products abroad via local sales representatives, and one company claims that its staff members speak a dozen languages, ranging from Urdu to Greek. But many U.S. photovoltaic systems are marketed via European intermediaries—not a good omen for the American industry.[45]

Government Programs and International Competition

Since the beginning of the space program, governments have also channeled funds into photovoltaics, mainly for research and development, but also to help commercialize technologies. Today the Western industrial countries and Japan spend about $150 million each year on photovoltaics, and most budgets are growing. Government support for photovoltaics deserves high marks for its technical achievements and for largely avoiding the politically-motivated boondoggles that have plagued so many government energy programs.[46]

The United States has traditionally had the largest government program in photovoltaics, but the tide has turned, with the Reagan administration slashing the photovoltaics budget from $150 million in

"One company claims that its staff
members speak a dozen languages, ranging from Urdu to Greek."

1980 to about $50 million in 1983. Meanwhile, Europe and Japan are
moving in the opposite direction. Solar cell budgets have increased in
several European countries, particularly in France and Italy. The
European Economic Community has also stepped up its support,
which includes a series of demonstration projects throughout
Europe. Together these programs have an annual budget ap-
proaching $50 million. In Japan the photovoltaics budget has risen
140 percent in two years to over $30 million. Including indirect gov-
ernment support, it is only slightly smaller than that of the U.S.
program.[47]

Governments do a number of things with their photovoltaics funds.
They help support R&D both in industry and university laboratories.
They use loans and grants to assist private companies in developing
low-cost manufacturing processes. And they create markets for solar
cells by purchasing photovoltaic systems for government facilities or
foreign aid projects. Effective support of the photovoltaics industry
requires a balance between these different activities, a point that
energy planners in Washington have recently ignored. The remaining
U.S. government funding is directed largely to basic R&D. Efforts to
help commercialize low-cost technologies have been gutted. Industry
representatives, noting the large investments required to bridge the
gap between the laboratory and marketplace, say the value of the
remaining R&D programs will be minimized if the government ig-
nores commercialization.[48]

Ironically, a conservative free-market government is ignoring the
signals of the marketplace, boosting support for the fading nuclear
industry while cutting off funds for rapidly growing solar cell companies. The results will not be apocalyptic, but they will slow progress
in the U.S. Fortunately, European and Japanese programs retain their
balanced approach. Perhaps the most effective government photovoltaics program in the world today is in Japan. Originally modeled
after the U.S. program, Japanese efforts today stress cooperation
between industry and government. The government helps private
companies take all technologies through the pilot plant stage, and
long-term programs guarantee a continuing commitment regardless
of year-to-year political changes.[49]

The close relationship between European and Japanese firms and their governments is one reason that they are so strong in international markets. Development assistance programs sponsored by France and West Germany help build markets for these nations' corporations, and photovoltaics has been a major beneficiary. In Japan, the Ministry of International Trade and Industry (MITI) has a worldwide reputation for its success in helping domestic corporations develop export markets. Though photovoltaics has received only modest support from MITI to date, assistance is expected to increase rapidly as Japanese firms become active in the international photovoltaics market, which until recently they saw as too small to be worth a major push. In contrast, the U.S. Agency for International Development has funded several demonstration projects, but does little to find markets for American technologies. The U.S. Government's Commerce Department and various overseas embassies today provide only limited, unfocused assistance for U.S. photovoltaics exporters.[50]

There is no easy way to predict future market shares in photovoltaics. However, those who do climb out on the forecasting limb are betting mainly on Japan. That country's recent success in marketing high-technology products around the world, combined with its growing national commitment to photovoltaics, make it a formidable competitor. In recent years Japanese officials have made it clear that they welcome the cuts in the U.S. photovoltaics program and hope to use this opportunity to close ground on the U.S. industry.[51] Unless U.S. Government programs are stepped up soon, or private industry dramatically boosts its investments, Japan will probably surpass the U.S. in solar cell sales by the close of this decade. Considering the strong interest of Japanese companies in export markets, American homes might one day be powered by solar electric systems made in Japan.

The photovoltaics industry nonetheless is likely to retain considerable geographical diversity, with American and European companies holding substantial market shares for the foreseeable future. Before too long, developing nations should also play an important part in the industry, first by assembling systems using photovoltaic cells made in industrial countries, and later manufacturing photovoltaics domestically. Already an American firm offers an "off-the-shelf" solar cell manufacturing plant for sale anywhere in the world—a quick way for a developing nation to enter the photovoltaics industry. Many developing countries will likely enter the photovoltaics business via joint ventures with established companies, and already several agreements have been signed and others are under negotiation. India is developing a largely indigenous photovoltaics industry using the skills of its scientific community and does not allow the import of photovoltaic systems from industrial countries. A surge in photovoltaics manufacturing in the Third World is likely within the next few years, led by Brazil, China, India and the Philippines.[52]

33

As the solar electricity market grows, the high cost of transporting bulky and breakable panels will likely encourage the manufacture of at least part of a solar electric system near where it will be used—just as happens with most home appliances today. Leaders of many Third World countries have indicated that they would like to develop indigenous photovoltaics industries, and some have already set up tariffs to restrict the import of complete systems. Indeed, joint ventures and licensing arrangements may become so widespread that the question of international leadership will be moot. In some cases, silicon might be refined in one country, the cells manufactured in a second and the system assembled in a third. At some point, most of the money spent on solar electric systems will likely be spent locally—for assembly and installation. The contrast with the world oil market and its thin line of tankers stretched across the oceans could hardly be greater.

Villages And Rooftops

No other energy technology has the versatility of photovoltaics. David Morris of the Institute for Local Self-Reliance in the United States observes that, "Using the same energy source—sunlight—and the same technology, we could have the most decentralized or the most centralized form of electricity generation in history."[53] The question of which uses for solar electricity will prove most economical and popular has already caused controversy. Advocates of centralized solar power cite the economic advantages of large photovoltaic systems since one set of power lines and control devices can serve a large installation. The partisans of decentralized solar electricity counter that if solar electric systems are placed on rooftops and fully integrated into industries and households, they could one day be cheaper than centrally generated power and permit energy independence as well. No clear resolution to this argument is in sight, and different uses of solar power are likely to emerge simultaneously.

The current market for solar cells is limited, focusing on specialized applications in areas not yet reached by utility lines. These remote-site uses are surprisingly numerous and diverse, however, spread throughout the developing and industrial worlds. And sales are growing rapidly. Portable solar cell modules, for example, are popular on boats and train cabooses. Remote mountain cabins and scientific research stations are beginning to rely on photovoltaics rather than diesel generators. A market for photovoltaics to power microwave repeaters and other communications installations is growing rapidly. Additional future uses for photovoltaics can be found at remote military installations—amounting to 100 megawatts of new solar cells each year in the United States alone, according to one study. In these and many other uses solar electric systems are now more economical and reliable than alternatives.[54]

One interesting new use for photovoltaics is the hand-held solar calculator, a device pioneered by Japanese companies. Using a tiny photovoltaic array that generates only a few thousandths of a watt, solar calculators have no battery and no on-off switch. When light strikes the solar cells the calculator automatically turns on. Because power requirements are so small, the solar array hardly adds to cost, and the calculators are sold competitively at $10 to $30. Without wires or batteries, solar calculators operate effectively with the light available in most rooms. And because both the calculator and its power system are solid-state devices, they should last indefinitely. The Japanese have developed a remarkably strong market that in just a few years has grown to between 10 and 20 million calculators annually. Industry observers believe that by 1985 half of the small calculators sold worldwide could be light-powered. Japanese companies are now exporting solar calculators and developing similar devices such as watches, children's toys and small battery chargers for use in remote areas.[55]

During the next decade, small stand-alone uses for photovoltaics will almost certainly predominate. There are still millions of houses in remote areas of Australia, Europe, North America and the Soviet Union that are not connected to power lines. And in developing countries, the majority of villages and rural areas still lack access to a steady supply of electricity. The dream of extending central grids into the "heart of darkness" has faded in the face of mounting Third World debts and the rising cost of vast networks of power lines. If millions of peasants and villagers are to have electricity in their lifetimes, most will have to rely on small, decentralized energy systems.[56]

Today the only electricity alternative in most villages is a diesel generator, typically run just a few hours a day to supply power for agricultural equipment and for a few lights in the evening. But diesel generators are expensive and—a bigger problem—unreliable. They require regular maintenance and an occasional complete overhaul. Since there are so few trained mechanics in rural areas, broken-down diesel generators are a common sight today throughout the Third World. A spare engine is normally recommended for diesel generating systems, to be used when the first breaks down. When maintenance and fuel costs are added to initial capital costs, diesel generated electricity costs between 20¢ and $1 per kilowatt hour—many times typical electricity prices in industrial countries.[57]

Several recent studies have reached similar conclusions about the cost of photovoltaics compared with diesel generating systems. For the small-scale uses for which diesels are inefficient and expensive, solar electric systems are already economical. Louis Rosenblum of the National Aeronautics and Space Adminstration concludes that the break-even point for replacing a diesel generator with a solar system is now between one and three kilowatts, depending on available sunlight and the exact cost of the photovoltaics. However, as solar cell prices fall 50 percent or more as expected in the next five years, solar power will become economical for virtually all electricity applications in Third World villages.[58]

Small-scale uses that are already economical include refrigerators for perishable food and medicines, communications systems, television sets, lights and mills. Water pumping is a particularly good use for solar electricity since sunlight is usually available when water is most needed and diesel pumps are oversized for many small farms and households. In the Third World, just a few hundred watts of power—miniscule by industrial nation standards—can have an enormous effect on village life, providing basic amenities for the first time.

Since the late seventies, a large part of the market for photovoltaics has been in developing countries. Interest has been particularly strong in the Philippines and Thailand, whose governments encourage the use of photovoltaics imported from Europe and the United States for communications, water pumping and other applications. A half dozen developing countries will soon have indigenous photovoltaics industries. In India and Pakistan there is a rapidly growing market for solar-powered micro pumps among small farmers. And the Pakistani government now plans to introduce solar electricity in 14 villages by 1984.[59]

The foreign aid programs of industrial nations have also begun to include photovoltaics, particularly those of France, the United States and West Germany. France has had major programs under way since the mid-seventies to install solar-powered pumps and television sets in West Africa. The television sets, modified to require as little as 20 watts of power, bring educational programs to people in remote areas at a reasonable cost. The U.S. Agency for International Development installed eight specially-designed solar-powered vaccine refrigerators at rural health centers in 1982 and has another dozen small-scale photovoltaics projects under way, mainly in Africa. In Somalia relief groups have brought in over a hundred photovoltaic water pumping systems to be used at refugee camps.[60]

International efforts to promote photovoltaics include studies and demonstration projects sponsored by the United Nations Develop-

"In the Third World, just a few hundred
watts of power—miniscule by industrial
nation standards—can have an enormous
effect on village life."

ment Programme (UNDP), the World Health Organization and the
World Bank. Their reports note that no established and reliable market now exists for photovoltaic systems in most developing countries
and that the quality of the equipment sold is uneven. In many areas,
protecting solar arrays from dust storms and keeping them clean is
also a problem. Development specialists, however, are generally confident that such obstacles can be overcome, and that photovoltaics
gradually will gain broad appeal for small-scale Third World uses.
They are particularly interested in the potential of solar-powered
pumping systems since the lack of fresh drinking water and adequate
irrigation is one of the most urgent needs in many rural areas. By
combining solar electric systems with efficient low-head pumps,
peasants and villagers should have an inexpensive and relatively
maintenance-free water supply.[61]

One of the more ambitious plans for photovoltaics is to provide
electricity for villages. The first experimental village system was installed on the Papago Indian Reservation in Arizona in 1978. Since
then its 3.5 kilowatts of cells have powered water pumps, lights,
refrigerators and communal washing and sewing machines for the
Schuchuli village's 95 residents, who previously had no electricity. In
1979 a 1.8-kilowatt photovoltaic array was installed at the Tangaye
village in Upper Volta by the U.S. Agency for International Development. It has been pumping 5,000 liters of water daily and grinding
enough grain for 640 families ever since. A much larger 25-kilowatt
system installed in a Tunisian village in 1982 is pumping water,
performing agricultural tasks and meeting household needs. The
village women, who once performed these tedious tasks by hand,
particularly welcome solar electricity.[62]

Several other experimental village systems have been set up and
carefully monitored, and all confirm the enormous potential of this
concept. For many of the world's poor who are still without electricity, photovoltaics may soon be a reliable power source. First, however, governments and international aid agencies must arrange innovative financing. One thing they can do is to use agricultural
extension services and subsidized loans that are already in place for
other technologies to encourage the use of solar pumps and mills.

Today only the wealthier developing countries are purchasing photovoltaics, while many African nations and others that most need a new source of electricity cannot afford it.[63]

Decentralized solar electricity systems are likely to play an important role in industrial regions as well. Solar cells can be placed on rooftops, turning buildings of all kinds into electricity generators. Photovoltaics thus has the potential to give ordinary people a measure of energy independence unprecedented in the modern world. With solar electric systems on their rooftops people will no longer be vulnerable to fuel price increases. With battery storage, power would stay on even if a storm knocks down local electrical lines. All of this brings psychological satisfaction as well as practical benefits. Together with wind turbines, geothermal energy, co-generation and small-scale hydropower, photovoltaics could gradually decentralize power systems. Utilities would not only produce power, they would become energy brokers, distributing electricity generated by tens of thousands of small producers.

If these new energy relationships are to evolve, rooftop solar power systems must be connected with the utility grids that are the main source of power in most industrial countries. The main reason for making the utility connection is economic: electrical battery storage can as much as double the total cost of solar power. But batteries are not needed if a utility provides backup electricity. This requires an electricity inverter, an inexpensive device that converts direct current into alternating current compatible with the electricity in utility lines. Special controls regulate the electricity going in and out of a building, and two meters are installed—one measuring the electricity sold by the utility to the household and the other measuring the power going from the house to the utility. Another possibility is a single meter running backward as well as forward. In some months, the utility might send the customer a check rather than a bill.[64]

Architects and engineers in the United States and Japan have already begun designing solar electric buildings. Like solar water heaters, photovoltaic systems require an unshaded southern exposure and durable materials. Lightweight photovoltaic panels need little additional structural support, but they do require more roof space than thermal collectors do—30- to 50-square meters for a typical household-size system in a sunny region.

In the United States approximately one-third of existing residential buildings could use photovoltaics—a total of 20 million houses. However, taking maximum advantage of photovoltaic electricity will require that electricity be used more efficiently and that new houses be designed with this technology in mind. A 1981 study by the Solar Energy Research Institute shows that in most parts of the United States a moderately-sized solar electric system on an energy-efficient house can provide *more* electricity than the houses uses. However, much of this power would be generated when it is not needed by the homeowner (and would be sold to the utility). The household would depend on the utility for electricity when the sun is not shining.[65]

The world's first specially-designed photovoltaic residence was built in Phoenix, Arizona in 1980 with the help of the U.S. Department of Energy funding. Since then, approximately a dozen solar electric buildings have been built, most of them experimental houses in the United States. At least one architectural firm has begun designing solar electric homes commercially, and others are considering entering the field. Right now, however, this is a specialty market for wealthy individuals with an adventurous spirit. A photovoltaic system today adds at least $60,000 to the cost of a house, and the solar electric homes sold so far have mainly been in the $200,000 to $400,000 range. These houses have nonetheless been an important proving ground, allowing architects to experiment with different designs, improving their aesthetics and ruggedness.[66]

Because rooftop solar systems are exposed to the elements and falling debris, they must be exceptionally durable. Recent innovative designs integrate solar cells into the roofs themselves, and solar cells may one day become a major roofing material. One design mounts photovoltaic panels on tracks, allowing them to slide easily on or off the roof. Another company has developed a photovoltaic shingle that provides power and also protects the roof. In all cases ease of installation and repair is essential. Some solar electric systems can be added to a properly designed house in just a few hours. Architects have already demonstrated that solar electric buildings can be attractive. While some are brashly modern—dominated by their photovoltaic systems—others have a variety of traditional styles that make the solar systems barely noticeable.[67]

A bona fide market for solar electric houses should emerge when photovoltaic module prices fall to about $2 per watt—approximately one-quarter their current level. This should happen by the early nineties. At this price a three-kilowatt photovoltaic system providing 50 percent of an energy-efficient household's electricity needs could be installed for between $12,000 and $18,000.[68] With tax credits such those already in place in some countries, solar electric houses would clearly be economical. And as prices fall below $2 per watt, solar electric homes will be generating electricity at a price competitive with utility generated power.

Some people look at rooftop photovoltaics as an energy panacea that will soon price utilities right out of the electricity business. Such views are fanciful and could be self-defeating. Solar electricity is not going to become cheap anytime in the foreseeable future. Rather, photovoltaics will gradually reach price competitiveness with conventional generating technologies and then only slowly undercut them, probably around the turn of the century.

Communities in all but the most overcast climates should eventually be able to obtain at least a quarter of their electricity from rooftop photovoltaics. But these photovoltaic systems will never stand alone if the goal is economical, uninterrupted power. The sun often shines brightly only a few hours a day, but electricity is needed continuously. It makes sense, therefore, to link solar electric systems with other generating technologies, including the many other renewable energy sources now being developed. Careful planning can produce a diverse but still economical and reliable power system.[69]

Solar Power Stations

Decentralized uses of photovoltaics notwithstanding, large, centralized photovoltaic power systems have long fascinated some scientists and planners. As early as the sixties University of Arizona researchers proposed the use of massive solar arrays placed in desert regions to supply much of the world's power.[70] Large solar electric systems do have some modest economies of scale, and, at least until recently, they seemed to match most closely the popular conception of how a modern power system should be structured. Utilities have a growing interest in photovoltaics, and centralized solar power, they feel, ensures that they will get a piece of the action.

41

The most dubious plan for harnessing solar electricity is the satellite solar power station. Several researchers in the United States have proposed placing large arrays of solar cells—about the dimensions of Manhattan—in stationary orbit around the earth, using microwave transmitters to convey the power to land-based receivers. A satellite power station, proponents point out, would receive intense 24-hour sunlight outside the earth's atmosphere. Designing and building the first such "Manhattan-in-the-sky" would take at least 20 years and cost well over $100 billion, making it the largest construction project ever undertaken. It would also be the most centralized form of electricity generation in history, but scientists agree that it is a technically feasible concept.[71]

Good theory does not always make good economics, and recent studies cast serious doubt on the satellite solar power idea. It will be decades at least before the vast quantity of materials needed to build an orbital power station can be economically boosted into space—or mined from the moon. Some skeptics question whether construction will ever be economical considering the large amount of energy required to overcome the earth's gravitational pull. More disturbing are the potential health and environmental effects of a high-energy beam aimed at the earth's surface. People would inevitably be exposed to low levels of microwave radiation that might cause health problems. The earth's atmosphere might also be damaged irreparably. Because of the combination of staggering economic costs and environmental risks, the U.S. National Academy of Sciences recommended in a 1981 study that no funds be committed to solar power satellites—an unusually firm judgment by the Academy against further research. An additional problem with satellite solar power is the aura of science fiction it gives to a technology with many here-and-now applications.[72]

Other centralized photovoltaic designs show more promise. Government and private researchers in the last few years have begun designing large, earth-based photovoltaic power systems. The effort, however, has been overshadowed by much larger programs in Europe and the United States to develop solar thermal power plants using concentrated sunlight to boil water and turn a turbine and generator.[73] Although these more conventional solar systems have a large potential in some sunny regions, photovoltaics likely will be more economical in the long run—and more flexible as well. Since 1980 government funding of solar thermal power systems has declined while funding of photovoltaics has risen in most countries.

Only a few large photovoltaic projects have been built so far. A 240-kilowatt solar power project using concentrators was completed in Arizona in 1982. It provides electricity for the Sky Harbor Airport near Phoenix and was largely financed by the U.S. Department of Energy. A similar system, funded jointly by the Saudi Arabian and U.S. governments, was installed in Saudi Arabia in 1982. It generates 350 kilowatts of solar power with diesel back-up for three Saudi villages. Perhaps most exciting is the 200-kilowatt photovoltaic powered solar cell manufacturing plant built by the Maryland-based Solarex Corporation in 1982. Solarex calls this $6 million facility, financed without government support, a "solar breeder" because its large bank of solar cells provides enough electricity to manufacture another 200 kilowatts of cells every 28 days. The system is electrically independent, with batteries for backup power. Solarex officials have begun selling modified versions of the plant to industries in developing countries lacking extensive utility grids.[74]

Meanwhile, utilities in California lead in integrating solar power systems into their electricity grids. The Sacramento Municipal Utility

District (SMUD) has begun work on a 100-megawatt (100,000 kilo-watt) photovoltaic power plant scheduled for completion by 1994 at a total cost of over $250 million. The first phase of the project, funded partly by the federal and state governments, will place one megawatt (1,000 kilowatts) in operation by 1984. The project will use flat plate arrays that occupy 1,100 acres of now-vacant land adjacent to the Rancho Seco nuclear power plant, and generate a tenth as much power as the nuclear plant does. The first privately funded photo-voltaic power plant is under way in southern California. Financed and built by the ARCO Solar Company, the project will feed one megawatt of power from flat plate solar arrays mounted on computer-controlled tracking devices into the lines of the Southern California Edison Company in 1983. In Italy a one-megawatt government-funded photovoltaic plant called the Delphos project is also sched-uled for completion in 1983.[75]

These utility projects are harbingers of things to come. As many as ten similar projects will likely be announced in the next few years, mostly in areas of the western United States and southern Europe with ample sunlight and existing oil-fired power plants that utilities wish to decommission. Utility planners point out that land avail-ability will not be a problem for the foreseeable future, particularly since there is much unused land adjacent to existing power plants and along power line rights of way. Beyond boosting electricity sup-plies, these projects create a large new market for the photovoltaics industry, helping bring down costs and improving the economic feasibility of all photovoltaic installations. The Sacramento project alone could, by the mid-eighties, add more solar cells each year than are currently produced worldwide.

The Future of Solar Power

The big news in the photovoltaics industry in the early eighties is that it is prospering—despite a worldwide recession in which many more established industries are on the brink of insolvency. Production of solar cells has expanded fivefold during a three-year period in which the world economy grew at a meager 1 percent annual rate. Based on manufacturing plants now under construction and unprecedented marketing efforts on the part of several large companies, prolific growth of about 50 percent per year is virtually guaranteed for the next three to five years. By 1986, the photovoltaics industry world-wide will probably be producing five times as many solar cells as it did in 1982.

Much more difficult to answer is the question of how widely photo-voltaics will be used in the more distant future. Since the early seventies, many predictions have been made about photovoltaics, but most should be taken with a grain of salt. They are usually based on dubious assumptions about the potential for cost reduction, future economic conditions and the rate of growth in electricity demand. As the estimates and assumptions are amassed, uncertainty multiplies.

The goal of the U.S. photovoltaics program, as formulated by Congress in 1978, was to double the manufacture of solar cells each year in order to reach an annual output of 2,000 megawatts of cells by 1988 (250 times the 1982 total). The U.S. Department of Energy subsequently established a target of obtaining one quad (quadrillion Btu's) of energy from photovoltaics by the year 2000. This would require an installed capacity of over 50,000 megawatts, which should supply about half as much power as nuclear plants provide today.[76]

The United States has already fallen short of the production trends required to meet these goals, and there is now general agreement that they will be reached much later than planned. The Reagan administration's cutbacks in the photovoltaics budget and its abandonment of commercialization programs have dealt U.S. photovoltaics goals an additional crippling blow.[77]

What went wrong? Does the slippage of these targets mean we should reduce efforts to harness photovoltaic electricity? In truth, the main problem is not with solar cell technology but with the goals themselves, which call for unrealistically rapid progress. Photovoltaics continues to follow a remarkably ambitious cost curve, but because of the large investment needed in automated manufacturing processes, progress gets tougher the further prices fall. Laboratory achievements

> "Energy conservation and modest eco-
> nomic growth are now the driving forces
> in the energy marketplace—something
> that no realistic photovoltaics forecast
> can ignore."

have been impressive, but time and money are needed to get from the lab to the factory, an inevitable truth in any new industry that photovoltaics forecasters tried to sidestep. Their goals called for solar electricity to become widespread much faster than has nuclear power, while spending only a small fraction of the funds that have been committed to nuclear power development.

45

Predictions about photovoltaics also often assume that the world has an insatiable appetite for electricity. Many analysts echo the unrealistic forecasts of the utility industry, which still sees electricity use in industrial countries growing by two-thirds in the next 18 years. The unspoken assumption in these forecasts is that despite rapidly rising electricity prices, the world will continue to substitute electricity for all other forms of energy. Some boost their forecasts even higher by assuming that the electric car will be commonplace within the next decade.[78]

Such analyses are well out of line with current energy trends. Energy conservation and modest economic growth are now the driving forces in the energy marketplace—something that no realistic photovoltaics forecast can ignore. Data from several major industrial countries show 1 to 2 percent yearly growth in electricity use in recent years, in some cases slower than the growth rate of the economy as a whole. Credible forecasters now project very slow or even no growth in electricity use during the next two decades. Interestingly, this low growth rate has been partly caused by the very trends that boost the prospects for photovoltaics—soaring costs for coal and nuclear power plants. Electricity prices will almost certainly continue to rise during the next decade, since coal and nuclear energy are by far the largest new power sources coming on line during this period, and many power plants are suffering massive cost overruns that have yet to be financed. Assuming rapid electricity growth during this period is wishful analysis reminiscent of the early days of nuclear power.[79]

Because of slower growth in electricity use, the absolute size of the market for photovoltaics during the next two decades will be smaller than was assumed a few years ago. In the major industrial countries, which already have considerable utility overcapacity, few new power plants are needed. New electricity generation, however, will be needed in developing countries and in industrial nations, as older power plants are replaced. These will be sizable markets for photovoltaics, though not as large as was once thought.

As a result, photovoltaics will likely be introduced gradually over the next 20 years. This is actually an asset, because if solar electric systems are introduced slowly, they can be linked effectively to existing power sources and employed in the most efficient manner. Nuclear power provides a good example of what can go wrong if a technology is too quickly forced on an unprepared world.

In an energy-conserving era, solar electricity has a major role to play. Photovoltaic technologies and energy conservation technologies have a naturally synergistic relationship. The lower the power requirements of a particular device, the more practical solar electric systems become. One improvement helps encourage another. French and Japanese companies, for instance, have worked hard to reduce the electricity needs of television sets and calculators so that they can be powered with solar cells. A large American "all-electric" house would be hard to power with photovoltaics, but the "low-energy" houses now popular in Scandinavia would be quite adaptable. The more that electricity requirements can be reduced in households and industries throughout the world, the closer we will be to widespread, practical use of photovoltaics. Photovoltaics, in turn, will spur the increased efficiency of motors and buildings.[80]

A few predictions are now possible. Based on the most recent cost trends, photovoltaics should begin replacing most diesel generators and become the largest new source of electricity for villages in developing countries within a few years. Third World governments and international aid agencies, however, must help introduce solar electric systems if they are to be widely used. Most important are financing mechanisms for poorer areas of the Third World. By the late eighties, centralized photovoltaic power stations should become conventional generating technologies in many regions, and utilities will help lead the way in introducing solar electricity. By the early nineties, rooftop solar electric systems should begin to catch on.

Worldwide annual production of photovoltaics is likely to rise to between 200 and 300 megawatts of capacity by 1990 and to over 1,000 megawatts by the end of the century. This will make the photovoltaics business a billion dollar industry by the late eighties and push it close to the $10 billion mark by the year 2000. Total installed solar electric capacity will probably be between 5,000 and 10,000 megawatts by century's end. Although this will be enough capacity to supply only a few percent of the world's electricity, it will set the stage for the rapid introduction of solar electricity in the following years. By the middle of the next century photovoltaics may be providing between 20 and 30 percent of the world's electricity and serving as a cornerstone of a sustainable global power system.[81]

The exact pace of photovoltaics development and the installed capacity in particular years will depend mainly on how fast large-scale, low-cost manufacturing processes are introduced. Today the largest photovoltaics plants being built produce between two and three megawatts worth of solar cells a year. Building more advanced plants that annually produce solar cells capable of generating 20 or 30 megawatts could lead to module costs of less than a quarter the current figure. The catch, of course, is that such plants will be expensive and companies will not build them until the market grows substantially.

The most important step that governments can take is to help stimulate the market for solar electric systems through the use of tax credits and loans, and by purchasing photovoltaics directly for use on government buildings, military facilities and for the poor. Demonstration projects and educational programs such as those of the European Economic Community can also encourage production. Equally critical is the success of the photovoltaics industry in raising needed capital. Careful planning and close relationships between photovoltaics companies and the investment community are essential.

If government support of photovoltaics is compared with support of nuclear power—another modern technology for electricity generation—the disparity is obvious. Even today governments spend several billion dollars annually on advanced nuclear technologies—

over ten times as much as they spend on photovoltaics. Yet even under optimistic assumptions breeder reactors and fusion power plants will take longer to enter widespread use than will photovoltaics.[82] While dozens of companies are working independently on photovoltaics, no private firm has invested a dime in breeder reactors without massive government backing. By restoring some balance to energy development programs, governments could get more for their taxpayers' money and dramatically boost the pace of photovoltaics commercialization. If solar cell technology had received the kind of intense multi-billion dollar support in the fifties that nuclear power was given, photovoltaic systems would already be common on rooftops.

Of course, the benefits of solar electricity cannot be measured simply in kilowatts or percentage points alone. Photovoltaic systems cause fewer environmental problems than any other means of generating electricity. They are a particularly striking contrast to coal and nuclear power, the most rapidly growing electricity sources today. Each of these poses environmental risks on an unprecedented global scale. Photovoltaics, on the other hand, is a major step toward an ecological and sustainable energy system.

Photovoltaics also has a strategic value that other energy sources lack. It will make small amounts of power available in virtually any corner of the globe, allowing developing countries to leapfrog the now prohibitively expensive process of extending electricity grids to all areas. Rarely do the world's poorest have a chance to benefit from one of the world's most advanced technologies. Photovoltaics not only can help remove some of the drudgery of village life, but by powering communications systems it can help educate people and bind together emerging nations.

The potential contribution of photovoltaics is perhaps most visible in the happiness of an African nomad who experiences running water for the first time. It is also clear in the improved health of an Indian villager who uses solar electricity to refrigerate food and medicine. Thanks to the advances being made in this seemingly exotic new technology, the living standards of hundreds of millions of people can be significantly raised in the next few decades.

1. Quotation from *ARCO Solar News*, April 1982.

2. Figure assumes a 3-kilowatt system at an installed cost of $17 per peak watt or close to twice the cost of the photovoltaic module alone. The cost could be much higher with a larger system. For details, see sections of this paper: "The Cost Reduction Game," and "Villages and Rooftops."

49

3. Projections are the author's and are described more fully in the last section of this paper: "The Future of Solar Power."

4. The early development of photovoltaic cells is described in Ken Butti and John Perlin, *A Golden Thread: 2500 Years of Solar Architecture and Technology* (New York: Van Nostrand Reinhold and Co., 1980), and Paul D. Maycock and Edward N. Stirewalt, *Photovoltaics: Sunlight to Electricity in One Step* (Andover, Mass.: Brick House Publishing, 1982). Additional details were supplied by Calvin Fuller, who was on the original Bell research team, private communication, March 1, 1982.

5. "Solar Cell is Ready for Commercial Jobs," *Business Week*, July 20, 1957.

6. Photovoltaics has had advocates in the scientific community since the fifties, but until recently these people lacked any real political clout. For a good history of U.S. attitudes toward solar energy and its political status vis-a-vis nuclear power, see Lamont C. Hempel, *The Politics of Sunshine*, an unpublished doctoral dissertation of the Public Policy Program, Claremont Graduate School, Claremont, Calif., 1982.

7. Cost estimate is from Calvin Fuller, private communication, March 1, 1982.

8. Maycock and Stirewalt, *Photovoltaics*.

9. *Ibid.*

10. The 8,000 kilowatts figure is the author's estimate based on the assessments of various industry analysts. See, for example, "IEEE Conference Report: Optimism Prevails," *Photovoltaic Insider's Report*, November 1982, which describes the results of a September survey of photovoltaics analysts. Historical figures are based on Strategies Unlimited, *1980-81 Market Review* (Mountain View, Calif.: 1981). The 1,800 houses figure is based on the assumption that the houses use an average of 800 kilowatt-hours of electricity each month and that the solar arrays operate on average at 25 percent of their capacity—typical of the southern U.S. Since electricity is often used at times when the sun is not shining, more photovoltaics and electrical storage would be required to actually meet the needs of this many houses.

11. Good overviews of conventional solar cell technology are included in Maycock and Stirewalt, *Photovoltaics,* Ed Roberton, ed., *The Solarex Guide to Solar Electricity* (Rockville, Md.: The Solarex Corporation, 1979), Bruce Chalmers, "The Photovoltaic Generation of Electricity," *Scientific American,* October 1976, and Yvonne Howell and David Adler, "How Silicon Solar Cells Work," *Sunworld,* Vol. 4, No. 1, 1980.

12. The photovoltaics industry is described in U.S. Department of Commerce, "Photovoltaics Industry Profile," unpublished, 1981, Barrett Stambler and Lyndon Stambler, *Competition in the Photovoltaics Industry: A Question of Balance* (Washington, D.C.: Center for Renewable Resources, 1982), and Science Applications, Inc., *Characterization and Assessment of Potential European and Japanese Competition in Photovoltaics* (Springfield, Va.: National Technical Information Service, 1979). Total sales and investment figures are not collected, so the figures used here are estimates based on the above sources and the assessments of industry analysts as reported in various issues of *Photovoltaic Insider's Report* and *World Solar Markets.*

13. Figures on current uses of photovoltaics are estimates by U.S. photovoltaics consultant Paul Maycock, private communication, March 8, 1982.

14. The three examples are drawn from *Arco Solar News,* various issues.

15. Photovoltaics technology is fundamentally different from all other technologies for electricity generation used today. It is the only technology that does not rely on a mechanical turbine and electrical generator. In both cost and lifetime terms photovoltaics has the potential to achieve performances beyond those of any conventional technologies. One of the central challenges comes in designing support equipment that is longlasting enough to allow photovoltaics to reach its potential.

16. Unless otherwise noted, price figures used here are in U.S. dollars per peak watt of generating capacity, which is the amount of power a solar module yields when placed in full sunlight at noon. This does not include the balance-of-system costs for the support structure, wiring, installation, etc., which add 50 to 100 percent to the price of a photovoltaic system today—an issue discussed at the end of this section. The $1,500 figure for a 100-watt light bulb assumes a total system cost of $15 per peak watt. Such a system would fully light the bulb only when the sun is shining brightly. The actual cost of electricity generated by a solar electric system varies widely, depending on how sunny the climate. A 1-kilowatt system might produce three kilowatt hours of electricity each day in one region (12 percent capacity factor) and six kilowatt hours in another (25 percent capacity factor), yielding a 100 percent difference in electricity costs. The historical price figures are estimates based on various sources, including Strategies Unlimited, *1980-81 Market Review*. These figures are in current dollars, so if the figures were adjusted to account for inflation, the price trends would be even more dramatic.

17. Overviews of photovoltaics R&D programs are included in Science Applications, Inc., *Characterization and Assessment of Potential European and Japanese Competition in Photovoltaics,* Robert R. Ferber, U.S. Jet Propulsion Laboratory, "The Status of Foreign Photovoltaics R&D," in U.S. Congress, House of Representatives, Subcommittees on Energy Development and Applications, and Investigations and Oversight, Committee on Science and Technology, Joint Hearings, June 3, 1982, and Monegon Ltd., *The Future of Photovoltaic Electricity: The Next 20 Years* (Gaithersburg, Md.: Monegon, Ltd., 1982).

18. Author's estimates based on Robert R. Ferber, private communication, November 11, 1982, and International Energy Agency, *Energy Research Development and Demonstration in the IEA Countries* (Paris: 1982).

19. Although photovoltaics and microelectronics are related, the cost reduction strategies employed are quite different. Microelectronics costs have been reduced primarily by squeezing an extraordinary amount of complicated electronic circuitry onto tiny silicon wafers, whereas photovoltaics researchers aim to produce large quantitites of a semiconductor material cheaply. Cost reduction in photovoltaics is mainly dependent on the development of inexpensive mass production techniques.

20. Techniques for reducing the cost of single-crystal silicon cells are described in Henry Kelly, "Photovoltaic Power Systems: A Tour Through the Alternatives," *Science,* February 10, 1978, Jeffrey L. Smith, "Photovoltaics," *Science,* June 26, 1981, Charles F. Gay, "Solar Cell Technology: An Assessment of the State of the Art," *Solar Engineering,* March 1980, and H. L. Durand, "Photovoltaics: Present Status and Future Prospects," *Sunworld,* Vol. 4, No. 1, 1980.

21. Companies that have developed processes to grow ribbons or sheets of single-crystal silicon cells include the Energy Materials Corporation, the Mobil Solar Energy Corporation, Motorola, Inc. and Westinghouse. All are in the

U.S. and some are licensing the technologies abroad. The various manufacturing processes are described in Maycock and Stirewalt, *Photovoltaics*, and Joseph L. Loferski, "Photovoltaics I: Solar Cell Arrays," *IEEE Spectrum*, February 1980.

22. Polycrystalline ingot technology is described in Maycock and Stirewalt, *Photovoltaics*, and J. Lindmayer and Z. Putney, "Semicrystalline Material from Metallurgical Grade Silicon," in *Fifteenth IEEE Photovoltaic Specialists Conference—1981 Proceedings*, Kissimee, Florida, May 12-15, 1981. Additional details were provided by Len Magid, Solarex Corporation, private communication, October 27, 1982. Solarex is the world leader in developing this technology. Crystal Systems in the U.S. and Wacker Chemie of West Germany are also working on polycrystalline cells.

23. J. Richard Burke, "Photovoltaics: Down to Earth at Last," *Solar Energy Research Journal*, Spring 1981. An overview of thin-film solar cell technologies is included in Maycock and Stirewalt, *Photovoltaics*.

24. Research progress in amorphous silicon is described in various papers included in *Fifteenth IEEE Photovoltaics Specialists Conference—1981 Proceedings*. See also "Sanyo Electric, Fuji Electric Boast Amorphous Si PV Efficiencies of 6.91%, 6.47%," *Solar Energy Intelligence Report*, April 27, 1982, Paul Danish, "Japanese Produce Amorphous Cells with Over 7.5% Conversion Efficiency," *Solar Times*, October 1981, "RCA Achieves Record 10% Efficiency for its Amorphous Silicon Cells," *Photovoltaic Insider's Report*, September 1982, and "Japanese Firm Gets 7% Efficiency," *Solar Engineering and Contracting*, October 1982. Japanese companies produced 700 kilowatts of amorphous silicon cells in 1982 and by 1984 will have at least 5,000 kilowatts of manufacturing capability, according to Paul Maycock, "My Fact Finding Tour of Japan," *Solar Age*, September 1982. Japanese companies developing amorphous silicon cells include the Fuji Electric Company, the Sanyo Electric Company and the Sharp Corporation. Leading U.S. firms include ARCO Solar, the Chronar Corporation, Energy Conversion Devices, IBM and RCA. A particularly exciting joint venture in amorphous silicon was announced by the Sharp Corporation in Japan and Energy Conversion Devices in the United States in 1982, described in Burt Solomon, "Sharp, ECD to Produce Silicon Cells in Japan," *Energy Daily*, June 22, 1982, and Wayne C. Johnson, "Sharp-ECD Solar Receives 3-MW/year PV Machine," *Solar Engineering and Contracting*, October 1982. The entry of the Chronar Corporation into the commercial market for amorphous silicon cells is described in Burt Solomon, "New Low-Cost Photovoltaic Cell Goes on Market," *Energy Daily*, June 3, 1982.

25. Research on thin-film materials is described in various papers included in *Fifteenth IEEE Photovoltaics Specialists Conference—1981 Proceedings.* See also Paul Blythe, Jr., "Thin Film Solar Cell Research Progresses," *Solar Engineering,* April 1981; "Boeing Achieves Highest Efficiency Ever for True Thin-Film Photovoltaic Cell," *Solar Energy Intelligence Report,* August 4, 1980; Solar Energy Research Institute, *Environmental Health, Safety, and Regulatory Review of Selected Photovoltaic Options* (Springfield, Va.: National Technical Information Service, 1982).

26. Donald L. Feucht, "Recent Progress in the Development of Advanced Solar Cells" and J. Stankiewicz and W. Giriat, "Materials for Tandem Solar Cells," both from *Fifteenth IEEE Photovoltaics Specialists Conference—1981 Proceedings.*

27. "LBL Develops Simple, Inexpensive Method for Dissociation of Water with Sunlight," *Solar Energy Intelligence Report,* September 27, 1982; "Texas A&M Hails Cheap Hydrogen," *The Energy Daily,* October 12, 1982.

28. Various concentrator designs are described in papers presented to the *Fifteenth IEEE Photovoltaics Specialists Conference—1981 Proceedings.* See also E. C. Boes, B. D. Shafer and D. G. Scheueler, "Economic Motivation for Photovoltaic Concentrator Technology," Sandia National Laboratory Report, unpublished, 1981.

29. A strong case for concentrator technology's future competitiveness in sunny regions is made in Harbinger Research Corporation, *Photovoltaic Power Systems Patents: A Technical and Economic Analysis* (White Plains, N.Y.: Madsen Russell Associates, Ltd., 1982).

30. U.S. Department of Energy, *Multi-Year Program Plan, National Photovoltaics Program* (Washington, D.C.: U.S. Government Printing Office, 1980).

31. Manufacturers often make claims about efficiencies and costs based on laboratory experiments. Such claims must be studied carefully in order to judge their significance. There is a large gap in this field between laboratory technologies and commercial technologies. Some companies have failed to live up to the short-term expectations fostered by their own executives. European and Japanese goals are described in EEC Commission, *Photovoltaic Power for Europe* (Dordrecht, Netherlands: D. Reidel Publishing, 1982), and Robert R. Ferber, "The Status of Foreign Photovoltaics R&D."

32. The information on balance-of-system costs is from Louis Rosenblum, "Cost of Photovoltaic Energy Systems as Determined by Balance-of-System

54 Costs," a National Aeronautics and Space Administration Technical Memorandum, June 1978. Updated figures, including the 80 percent projection for the early nineties, were provided by Louis Rosenblum, a photovoltaics consultant, private communication, November 4, 1982. Further analysis of balance-of-system costs can be found in the Electric Power Research Institute, *Photovoltaic Balance-of-System Assessment* (Palo Alto, Calif.: 1982).

33. Louis Rosenblum, "Cost of Photovoltaic Energy Systems as Determined by Balance-of-System Costs." Battery storage is discussed further in William D. Metz, "Energy Storage and Solar Power: An Exaggerated Problem," *Science*, June 1978.

34. These figures are the author's estimates based on assessments of various analysts. See, for example, "Maycock Predicts PV Future," *World Solar Markets*, August 1982, and "IEEE Conference Report," *Photovoltaic Insider's Report*, November 1982.

35. Figures are estimates based on discussions with various industry analysts.

36. The early development of the photovoltaics industry is discussed in Maycock and Stirewalt, *Photovoltaics*.

37. Oil company involvement in the photovoltaics industry is described in detail in Stambler and Stambler, *Competition in the Photovoltaics Industry*. Major oil companies that have invested in photovoltaics include Amoco, ARCO, Chevron, Elf Aquitaire, Exxon, Gulf Oil, Mobil Oil, Phillips Petroleum, Royal Dutch Shell, Shell U.S., Sohio and Total.

38. Ray Reece, *The Sun Betrayed: A Report on the Corporate Seizure of U.S. Solar Energy Development* (Boston: South End Press, 1979); Stambler and Stambler, *Competition in the Photovoltaics Industry*. Karl Frieden is quoted in Ralph Flood, "Big Oil Reaches for the Sun," *New Scientist*, November 12, 1981.

39. George Tenet is quoted in "SEIA Hits CRR Report on PV Market, Calls Industry 'Extremely Competitive'," *Solar Energy Intelligence Report*, March 1, 1982. Morris Adelman is quoted in "Big Oil Reaches for the Sun."

40. These points are discussed extensively in Stambler and Stambler, *Competition in the Photovoltaics Industry*.

41. Ronald L. Wilcox, in "Oil Companies and Photovoltaics: A Potential

Monopoly?," *Solar Law Reporter*, November/December 1981, points out that existing U.S. anti-trust laws will not impede oil company purchases of photovoltaics firms. Other approaches, including the granting of small business loans, are therefore needed to maintain a diverse industry.

42. The international photovoltaics industry is described in Stambler and Stambler, *Competition in the Photovoltaics Industry*, Science Applications, Inc., *Characterization and Assessment of Potential European and Japanese Competition*, and William J. Murray, "Marketing Photovoltaics: The Europeans are Coming," *Solar Engineering*. Additional information was provided by Robert R. Ferber, U.S. Jet Propulsion Laboratory, private communication, November 11, 1982 and Robert Johnson, Strategies Unlimited, private communication, November 12, 1982.

43. Murray, "Marketing Photovoltaics"; "France Plans PV Expansion," *World Solar Markets*, July 1982; "Will German Solar Survive the Recession?" *World Solar Markets*, month unknown, 1981; Douglas L. Finch, "The Japanese Photovoltaic Threat," *Solar Age*, February 1981; Maycock, "My Fact Finding Tour of Japan."

44. Solarex joint ventures are described in the *Solarex International Newsletter*, (various issues). Several of the large Japanese photovoltaics companies have agreements with U.S. companies for the transfer of solar cell technologies, according to news reports.

45. The natural advantages of European and Japanese companies in Third World markets are well-described in Science Applications, Inc., *Characterization and Assessment of Potential European and Japanese Competition*. The current market in developing companies, which is particularly competitive in Southeast Asia, was described by Louis Rosenblum, a photovoltaics consultant, private communication, November 4, 1982.

46. Government programs are described in William D. Metz and Allen L. Hammond, *Solar Energy in America* (Washington, D.C.: American Association for the Advancement of Science, 1978), James W. Fortenberry and John L. Hesse, "Overview of the National Photovoltaics Program," in *Proceedings of the 5th National Passive Solar Conference*, American Section of the International Solar Energy Society, Amherst, Mass., October 19-26, 1980, Paul Maycock, "Overview of the U.S. Photovoltaic Program," in *Fifteenth IEEE Photovoltaic Specialists Conference—1981 Proceedings*, R. Ferber and K. Shimada, "Japanese Photovoltaic R&D," U.S. Jet Propulsion Laboratory, Pasadena, Calif., unpublished, 1982, New Energy Development Organization, "R&D of Photovoltaic Conversion Systems," Government of Japan, Tokyo, unpublished, 1980, "EEC PV Conference Emphasizes Research," *World Solar Markets,* June 1982, and Science Applications, Inc., *Characterization and Assessment of Potential European and Japanese Competition.*

47. Robert R. Ferber, "The Status of Foreign Photovoltaics R&D"; "EEC PV Conference Emphasizes Research." The U.S. photovoltaics budget in fiscal year 1982 was $78 million. The budget for fiscal year 1983, which began in October 1982, has not been decided on by Congress, but most likely will be between $45 million and $50 million, according to Dick Munson of the Solar Lobby, private communication, November 8, 1982.

48. These points are made in private by both industry and government researchers, though many will not openly criticize the Reagan approach for fear of losing government contracts or their jobs. An additional problem plaguing the U.S. R&D program today is uncertainty. Recent budget appropriations by Congress have been made after a fiscal year begins, so researchers do not know how much their programs will have to be cut even as those programs proceed. This has caused chaos for some program managers and has led to a general decline in morale among researchers. For further industry views on the effects of budget reductions, see U.S. Congress, General Accounting Office, *Industry Views on the Ability of the U.S. Photovoltaics Industry to Compete in Foreign Markets* (Washington, D.C.: September 1981).

49. The differences between the various national R&D programs are described in Robert R. Ferber, "The Status of Foreign Photovoltaics R&D."

50. Science Applications, Inc., *Characterization and Assessment of Potential European and Japanese Competition;* U.S. Congress, General Accounting Office, *Industry Views.*

51. Dr. Takashi Horigome, Director of Solar Energy for the New Energy Development Organization in Japan, expressed satisfaction at the prospect of U.S. budget cutbacks in a private communication, June 6, 1981.

52. "Spire Offers a First: 'Off-the-Shelf' PV Cell, Module Integrated Production Lines," *Photovoltaic Insider's Report,* April 1982. Information on Third World companies is from Robert Johnson, Strategies Unlimited, private communication, November 12, 1982.

53. David Morris, *Self-Reliant Cities* (San Francisco: Sierra Club Books, 1982).

54. Remote-site uses for photovoltaics are described in Maycock and Stirewalt, *Photovoltaics* and in various periodicals.

55. "Japan Dominates PV Calculator Market," *World Solar Markets,* September 1981; Paul Maycock, "My Fact-Finding Tour," Robert Feldman, "Japan Leaps Ahead in Solar Energy: Opens Largest Solar Cell Plant," *Electronic Engineering Times,* September 14, 1981.

56. Information on the energy dilemmas facing developing countries can be found in the World Bank, *Energy in the Developing Countries* (Washington, D.C.: 1980), Thomas Hoffman and Brian Johnson, *The World Energy Triangle: A Strategy for Cooperation* (Cambridge, Mass.: Ballinger Publishing, 1981), and Amulya Kumar N. Reddy, "Alternative Energy Policies for Developing Countries: A Case Study of India," in Robert A. Bohm et al., eds., *World Energy Production and Productivity, Proceedings of the International Energy Symposium I* (Cambridge, Mass.: Ballinger Publishing, 1981).

57. Studies that have examined the economics and reliability of diesel generators include "Report of the Technical Panel on Solar Energy," prepared for the United Nations Conference on New and Renewable Sources of Energy, Nairobi, Kenya, August 10-21, 1981, Development Sciences, Inc., "The Economic Costs of Renewable Energy," Agency for International Development, unpublished, April 2, 1981, Louis Rosenblum, "Status of Photovoltaic Systems for Applications in Developing Countries," National Aeronautics and Space Administration, unpublished, 1981, and Norman L. Brown and James W. Howe, "Solar Energy for Village Development," *Science,* February 10, 1978. Although diesel engines are widely used in rural areas, there is a dearth of detailed surveys indicating their numbers and reliability. Anecdotal evidence provided by individuals who have traveled widely in developing countries indicates that the machines are out of order more often than not. This issue deserves study since if initial indications are right, diesel engines should be introduced more carefully and in many cases be replaced by renewable energy technologies.

58. Louis Rosenblum, "Status of Photovoltaic Systems"; Dennis Elwell, "Solar Electricity Generation in Developing Countries," *Mazingira,* Vol. 5/3, 1981; Lloyd O. Herwig, "Economic Comparisons Among Selected Operating Small Solar Power Systems," U.S. Department of Energy, unpublished, July 1981; Steve Hogan, *The Future of Photovoltaic Energy Conversion in Developing Countries* (Golden, Colo.: Solar Energy Research Institute, April 1980); Charles Drucker, "Third World Briefing: Photovoltaics Debated," *Soft Energy Notes,* May/June 1982. In some areas there are other renewable sources of electricity that may be more economical, such as wind power or biomass. However, these energy sources are scarce in many regions, and as the price of photovoltaics falls, it is likely to become the most widely used rural electricity technology. Sunlight is fortunately quite plentiful in most equatorial regions.

59. "PV Equipment Manufacturers See Growing Market in Third World," *Solar Energy Intelligence Report*, February 22, 1980; Rebecca Kauffman, "India Promotes Local PV for Space and Pumping," *Renewable Energy News*, April 1982; "Pakistan Gets First of 14 PV Generators," *World Solar Markets*, October 1981; Robert Gibbs, "Photovoltaic Cells: Transforming the Sun's Rays into Electricity," *R&D Mexico*, May 1981; Stephen Allison, "Selling Solar Pumps in Pakistan," *VITA News*, October 1982. Information on the market in the Philippines and Thailand was provided by Louis Rosenblum, private communication, November 4, 1982.

60. "NASA Approves Third World PV Projects," *World Solar Markets*, December 1981; Christopher Joyce and William G. Mahoney, "NASA Backs Worldwide Solar Power Project to Aid Poor," *World Environment Report*, March 15, 1982. Additional information on U.S. AID funded programs was provided by Louis Rosenblum, private communication, November 4, 1981. European aid programs and their promotion of photovoltaics are described in Science Applications, Inc., *Characterization and Assessment of Potential European and Japanese Competition*. The French program to introduce solar-powered television sets was described by Emile Gouriou of Bureau Yves Houssin, private communication, September 3, 1981. The 100 pumps in Somalia are described in *ARCO Solar News*, Summer 1982.

61. A major study on solar-powered pumps sponsored by the United Nations Development Programme and the World Bank and carried out by British engineering consultants Sir William Halcrow and the Intermediate Technology Development Group is described in "Solar Pumps Can Compete," *World Solar Markets*, August 1981. See also Jack J. Fritz, "Photovoltaic Water Pumping: A State-of-the-Art Review," National Academy of Sciences, unpublished, May 1982.

62. Bill D'Alessandro, "Villagers Light the Way: Solar Cell Power in Gunsight, Arizona," *Solar Age*, May 1979; William J. Bifano et al., "A Photovoltaic Power System in the Remote African Village of Tangaye, Upper Volta," NASA Technical Memorandum, unpublished, December 1979; "Ambitious 30 KW/p PV System to be Completed in Tunisian Village," *Photovoltaic Insider's Report*, September 1982; "100 KW/p Village Power PV System Being

Constructed on Greek Island of Kythnos," *Photovoltaic Insider's Report*, September 1982.

63. The immense energy financing problems facing developing countries and ways to resolve them are described in Hoffman and Johnson, *The World Energy Triangle*, The World Bank, *Energy in the Developing Countries*, and Maurice Strong and Mahbub ul Haq, *The Castel Gandolfo Report on Renewable Energy: Policies and Options*, presented to the North-South Roundtable Seminar at the United Nations Conference on New and Renewable Sources of Energy, Nairobi, Kenya, August 10-21, 1981.

64. Charles H. Cox III, "Power-Producing Homes: Making the Utility Connection," *Solar Age*, December 1981; Miles C. Russell, "An Apprentice's Guide to Photovoltaics," *Solar Age*, July 1981; Richard Kemp, *Solar Electricity—Space Age Technology on a Do-It-Yourself Level* (Ann Arbor, Mich.: Aatec Publications, 1981).

65. "Residential Retrofits Offer Enormous Existing Market, New Study Demonstrates," *Photovoltaic Insider's Report*, May 1982. Solar Energy Research Institute, *A New Prosperity: Building a Sustainable Energy Future* (Andover, Mass.: Brick House Publishing, 1981). The study examined the impact of a 4.5-kilowatt (35-square-meter) photovoltaic system installed on an energy-efficient house that uses between 420 and 480 kilowatt-hours of electricity monthly in three cities: Fresno, California, Washington, D.C., and Madison, Wisconsin. In Fresno and Washington there is a net sale of electricity by the homeowner to the utility, and in Madison they break even. However, in all cases the homeowner must rely on the utility for a sizable amount of power—between 46 percent and 62 percent of total needs.

66. Various solar electric houses are described in Burt E. Nichols and Steven J. Strong, "The Carlisle House: Solar Electric Residence is Energy Self-Sufficient," *Solar Engineering*, April 1982, Roger J. Osten, Jr. and Steven J. Strong, "A Prototype Passive Solar Heated and Cooled Photovoltaic Powered Residence for the Northeast," in *Proceedings of the 5th National Passive Solar Conference of the American Section of the International Solar Energy Society*, Amherst, Mass., October 19-26, 1980, and Dave McNary, "A Small Boom in Home Systems," *Solar Age*, August 1981.

67. Murray A. Toback, "PV Home Installation Yields Important Data," *Solar Engineering*, November 1981; Miles C. Russell, "Residential Photovoltaic Systems Designs," *Solar Engineering*, November 1981; Gordon F. Tully, J. Stewart Roberts, and Thomas A. Downer, "The Design Tradeoff for the Mid- to

59

Late-1980's: Photovoltaics versus Passive," *Proceedings of the 5th National Passive Solar Conference.*

60

68. These figures include balance-of-system costs of between $4 and $6 per peak watt. A 3-kilowatt system could supply more than half of electricity requirements in sunny areas and considerably less in some cloudy regions. Climate will continue to be an important determinant of economic feasibility.

69. For a discussion of diverse renewable resource-based electricity systems, see Daniel Deudney and Christopher Flavin, *Renewable Energy: The Power to Choose.* New York: W. W. Norton & Co., forthcoming. More work is needed on the integration of decentralized photovoltaic systems into utility grids. Preliminary studies indicate that some special measures will have to be taken, but that there are no overwhelming obstacles so long as the solar electric share of total power supply does not exceed 30 percent. Solar electricity will displace baseload generating plants in most regions and will reduce peak power requirements even further in hot, dry climates. For further information see M. Hassan and J. Klein, *Distributed Photovoltaic Systems: Utility Interface Issues and Their Present Status* (Washington, D.C.: U.S. Department of Energy, 1981), R. O. Mueller, B. K. Cha, R. F. Giese, "Solar Photovoltaic Power Systems: Will They Reduce Utility Peaking Requirements?", *Science,* July 10, 1981, and R. W. Boardman et al, *Impact of Dispersed Solar and Wind Systems on Electric Distribution Planning and Operation* (Springfield, Va.: National Technical Information Service, 1981).

70. A centralized, desert-based solar power system for the U.S. was proposed in Aden Baker Meinel and Marjorie Pettit Meinel, *Power for the People* (Tucson, Ariz.: privately published, 1970). A similar scheme was proposed by William Cherry of NASA, one of the early photovoltaics advocates. His plan, described in Maycock and Stirewalt, *Photovoltaics,* calls for solar arrays by 1990 to cover 31,500-square miles, or 1 percent, of U.S. land in order to supply all electricity.

71. The original proposal for a satellite solar power system is Peter E. Glaser, "Power from the Sun: Its Future," *Science,* November 22, 1968. Also see Peter E. Glaser, "Evolution of the Solar Power Satellite Concept: The Utilization of Energy from Space," presented at the Second United Nations Conference on the Exploration and Peaceful Uses of Outer Space, in Vienna, Austria, August 9-21, 1982. For a critical view of the idea, see Adam Hochschild, "Shuttling Manhattans to the Sky," *Mother Jones,* May 1978.

72. It would take 15 years for a satellite solar power system to generate as much energy as is required to build it, a very long energy payback period compared to most energy sources, including terrestrial photovoltaics, according to R. A. Herendeen, T. Kary, and J. Rebitzer, "Energy Analysis of the Solar Power Satellite," *Science,* August 3, 1979. Detailed critical assessments of the concept are National Science Foundation, *Electric Power from Orbit: A Critique of a Satellite Power System* (Washington, D.C.: 1981), and U.S. Congress, Office of Technology Assessment, *Solar Power Satellites* (Washington, D.C.: 1981). These studies point out that photovoltaics will be economical for use on earth long before they will be economical through orbiting power stations.

73. Extensive development of solar thermal power plants was the centerpiece of American and European solar research programs throughout most of the seventies. Several experimental plants have been erected in the southwestern United States, Italy and Spain. The largest, a 10-megawatt plant in Barstow, California, cost $140 million. These projects have been widely criticized, mainly because they are expensive and appear to draw funds away from other solar energy technologies. In the last few years solar thermal power plant budgets have been reduced and photovoltaics budgets increased in most countries. See Gerald W. Braun, *The U.S. Department of Energy Solar Thermal Systems Program* (Golden, Colo.: Solar Energy Research Institute, 1980), Clifford S. Selvage, "Some European Solar Thermal Power Systems," *Solar Engineering,* July 1979, and Brian Gallagher, "Solar Thermal Power Towers: A Costly Boondoggle?," Report Series No. 107, Citizens Energy Project, Washington, D.C., 1981.

74. Large photovoltaic projects are described in "Major Solar Projects Round the World," *World Solar Markets,* December 1981, and "Contractors for EEC Photovoltaic Generator Programme," *World Solar Markets,* month unknown, 1981. See also "Fresnel Lens PV Systems Starting Up at Saudi Villages, Phoenix Airport," *Solar Energy Intelligence Report,* April 5, 1982, J. J. Fendley, "Solar Energy in Saudi Arabia: The Soleras Project," *Helios,* July 1982, and "The World's First Solar Breeder," a 1982 promotional brochure of the Solarex Corporation. Several large photovoltaic projects sponsored by the European Economic Commission will be completed in 1983.

75. "ARCO to Build 1-MWe PV Array, Sell Power to SoCal Ed," *Solar Energy Intelligence Report,* April 5, 1982, and "100 MW—Phase One," *Renewable Energy News,* February 1982. The Italian project is described in "Major Solar Projects."

76. U.S. Congress, "Solar Photovoltaic Energy Research, Development, and Demonstration Act of 1978," Washington, D.C., 1978; U.S. Department of Energy, *Multi-Year Program Plan, National Photovoltaics Program*. A "maximum practical" figure of one quad for the year 2000 was used by the landmark study, *Domestic Policy Review of Solar Energy* (Washington, D.C.: U.S. Department of Energy, 1979). Since nuclear power plants operate on average at over 50 percent of their rated capacity and photovoltaics operate at under 25 percent, it takes at least two megawatts of solar cells to generate as much power as one megawatt of nuclear capacity.

77. U.S. Congress, General Accounting Office, *Probable Impacts of Budget Reductions on the Development and Use of Photovoltaic Energy Systems* (Washington, D.C.: March 1982).

78. For a utility industry view of electricity trends, see "33rd Annual Electrical Industry Forecast," *Electrical World*, September 1982. Even *Electrical World* has been greatly reducing its forecasts in the last several years as conservation has taken hold. For equally bullish views of electricity trends by photovoltaics analysts, see Monegon, Ltd., *The Future of Photovoltaic Solar Electricity: The Next 20 Years* (Gaithersburg, Md.: 1982), and Maycock and Stirewalt, *Photovoltaics*.

79. Growth in U.S. electricity use in the last five years has averaged 1.7 percent per year according to the U.S. Department of Energy, *Monthly Energy Review*, October 1982. Similar trends in Europe are reported in "EEC Nuclear Makes Big Gains Despite Overall Energy Decline," *European Energy Report*, February 5, 1982. Projections of very slow growth in electricity use include Solar Energy Research Institute, *A New Prosperity*, and Robert H. Williams, *Our Energy: Regaining Control* (New York: McGraw-Hill, 1981).

80. French efforts to lower the electricity requirements of television sets in order to power them with photovoltaics were described by Emile Gouriou of Bureau Yves Houssin, private communication, September 3, 1981. Many pioneering users of photovoltaic systems are working to reduce the power needs of appliances. See, for example, W. L. Peters, "Low Power Appliances Render Photovoltaics Affordable," *Canadian Renewable Energy News*, December 1981.

81. These projections are the author's and are based on the analysis in earlier sections of the paper, particularly the cost projections made in the section, "The Cost Reduction Game." The 20 to 30 percent limit is determined by space limitations in some regions, but more importantly by reliability problems that result if photovoltaics supplies too large a share of the electricity in a power grid. For a review of photovoltaics projections by various industry analysts see "IEEE PV Conference Report: Optimism Prevails," *Photovoltaic Insider's Report*, November 1982.

82. The Ford Foundation, *Nuclear Power: Issues and Choices* (Cambridge, Mass.: Ballinger Publishing, 1980); U.S. National Academy of Sciences, *Energy in Transition: 1985-2010* (San Francisco: W. H. Freeman and Co., 1979). Each of these studies takes an optimistic view of advanced nuclear technologies but concludes that neither breeder plants nor fusion reactors will be widely commercialized until well after the turn of the century.

CHRISTOPHER FLAVIN is a Senior Researcher with Worldwatch Institute and coauthor of *Renewable Energy: The Power to Choose* (W. W. Norton, Spring 1983). His research deals with renewable energy technologies and policies. He is a graduate of Williams College, where he studied Economics and Biology and participated in the Environmental Studies Program.